No Bicycle, No Bus, No Job

No Bicycle, No Bus, No Job

The Making of Workers' Mobility in the Netherlands, 1920-1990

Patrick Bek

Routledge
Taylor & Francis Group
LONDON AND NEW YORK

The publication of this book was made possible through the generous support of Rijkswaterstaat, Eindhoven University of Technology's School of Innovation Sciences, J.E. Jurriaanse Stichting, Stichting Unger-Van Brero Fonds, and the Foundation for the History of Technology.

First published in 2022 by Amsterdam University Press Ltd.

Published 2025 by Routledge
4 Park Square, Milton Park, Abingdon, Oxon OX14 4RN
605 Third Avenue, New York, NY 10158

Routledge is an imprint of the Taylor & Francis Group, an informa business

ISBN: 9789463723183 (hbk)
ISBN: 9781041183587 (pbk)
ISBN: 9781003700678 (ebk)
NUR 903

Cover photo: Workers leaving Philips factories by bicycle and bus; Eindhoven, 1958
Source: Philips Company Archives

Cover design: Coördesign, Leiden

DOI: 10.5117/9789463723183

For Product Safety Concerns and Information please contact our EU representative:
GPSR@taylorandfrancis.com
Taylor & Francis Verlag GmbH, Kaufingerstraße 24, 80331 München, Germany

Acknowledgements

The road to *No Bicycle, No Bus, No Job* has been as exciting as it was challenging. In researching archives scattered across the Netherlands, searching for untold stories in numerous historical collections, and writing in accessible prose, I had to overcome many hurdles and navigate my way along unfamiliar paths, sometimes not knowing what lay ahead. I could not have done this without the trust and encouragement of others.

I feel privileged to have worked on the Bicycle Challenges project at Eindhoven University of Technology. Without Ruth Oldenziel's unbounded enthusiasm, this project would not have happened. As my promotor, Ruth gave me the opportunity to learn, grow, and follow my gut feeling when choosing directions. My co-promotor Karena Kalmbach was always there to listen and give constructive feedback, guiding me through the academic mores. I want to thank Ruth and Karena for giving me the space and confidence to find my voice while writing this book.

I owe a special debt to project funders PON Holding, Rijkswaterstaat, and TU Eindhoven. Special thanks also to the faculty board of Industrial Engineering and Innovation Sciences for giving me extra time to finish this book, and Rijkswaterstaat, J.E. Jurriaanse Stichting, and Stichting Unger-van Brero Fonds for their generous subsidy to make this publication possible.

I am very grateful to Dick van Lente, who referred me to the Eindhoven history group. Many thanks to Frank Schipper, who helped me land in Eindhoven and organized inspiring bootcamps to get our working group going in the first months. I would like to thank Matthew Bruno, Henk-Jan Dekker, George Liu, Brett Petzer, Jan Ploeger, and Sun Qi for being a splendid working group. They provided important commentary on early drafts. Henk-Jan and Jan also generously shared some rich historical sources. Similarly, I thank History Lab colleagues Eric Berkers, Mila Davids, Jan Korsten, Harry Lintsen, Frank Veraart, Ginevra Sanvitale, and Erik van der Vleuten. I hold dear memories of visiting conferences with this group—moments like a pivotal conversation with Frank Veraart in Athens 2017 and desperately looking for a place to eat in Saint-Étienne with Henk-Jan and Ginevra, navigating crowded streets after France won the 2018 World Cup. I thank my colleagues in the Technology, Innovation, and Society group for providing a pleasant working environment, in particular Aleid, Aziiz, Darja, Edgar, Floor, Letty, Luc, Natascha, Peter, Sonia, and Tanja (Prague buddies!).

I would like to thank all those who offered their advice—practical, historical, and theoretical—during the writing of this book. I am indebted to doctoral committee members Adri Albert de la Bruhèze, Greet De Block, Simon Gunn, Karin Hofmeester, and Hans Jeekel for their constructive comments. Furthermore, my thanks go to Enne de Boer, Angela van der Kloof, Dick van Lente, Arwen Mohun, Peter Norton, Corjan van der Peet, Cemile Sezer, Hugo van der Steenhoven, Bert Toussaint, Pauliene van Vliet, and my peers at Tensions of Europe, ICOHTEC, CPSUM, WTMC, and T2M conferences and workshops for the many invaluable conversations we had. Similarly, a big thankyou to Leo Lucassen for accepting me as visiting researcher at the International Institute for Social History in 2019, enabling me to delve deeper into labor history and IISH collections.

Historical research is built on scarce and often hard to find historical source material. The work of archivists is indispensable to any historian. In particular, I thank Marc Rensen and Marianka Louwers of Philips Company Archives, and Tineke van Aalst of Tata Steel Central Archives for finding some of the great sources featured in this book. Writing an accessible book is yet another craft. I want to thank Ruth Oldenziel and Val Kidd for extensive text editing. Without their sense of language, eye for detail, and many hours of dedicated editing, this book would not have achieved its current form.

Last but not least, I would like to thank my family and friends: My parents for encouraging my childhood fascination with history, enabling me to pursue my interests, and supporting me in later career choices. Fritz and Brigitte for listening and putting work into perspective. My dear friends—Davey, Doris, Mark, Nils, Rens, and Tim in particular—for their encouragement and providing the necessary fun distractions from work. Many thanks also to Art and Marjon for letting me use their garden house for writing during COVID-lockdowns.

Above all I am grateful to my partner Sanne. Our boys Mats and Oskar came into our lives during the writing of this book, bringing us much joy, but at times also challenging our resilience, especially during lockdowns. Sanne agreed with me to embark on this project while we were enjoying our 2016 babymoon in Antwerp. She has supported me all the way, even when writing took up lots of my headspace. I could not have done this without her.

Rotterdam, 30 October 2021

Table of Contents

Figures

Introduction

In 1973, Dutch carpenter K. could not travel to his new job. For working people like him, the time and cost of getting to work are a crucial aspect of daily life. According to mobility historians, for most people, the opportunity to travel increased over the course of the twentieth century. This was not the case for everyone. International scholarship since the late 1960s has shown that the absence of affordable housing near work locations combined with a lack of safe, efficient, and affordable mobility options aggravate social exclusion for some. From this perspective, leading mobility researchers call for studying—but have yet to detail—how (uneven) power relations have historically enabled or inhibited people's mobility.[1] Historians have not followed up this call. While labor historians have a long tradition of analyzing power in relation to blue-collar workers' physical movements within factories and affordable nearby company housing, they have not studied in-depth the everyday commute. *No Bicycle, No Bus, No Job* redresses these omissions by researching how workers' mobility and job accessibility changed over time, and who contributed to this change in twentieth century Netherlands.

The case of K. illustrates how mobility was—and continues to be—an important resource for workers to capitalize on opportunities in modern liberal societies, built around the expectations of self-reliant and highly mobile citizens. Amid the recession following the 1973 oil crisis and global wave of deindustrialization, socialist newspaper *Het Parool* (1975) reported that 22-year-old carpenter K. was on trial for refusing what authorities deemed "suitable work." In light of layoffs, he had reported to the Regional Employment Office (Gewestelijk Arbeidsbureau, GAB), but the job offer meant traveling 11 km to work, a distance he considered too great: K. had neither a car nor access to public transit. He did own a bicycle, which was so old he could not

1 Tim Cresswell, "Towards a Politics of Mobility," *Environment and Planning D: Society and Space* 28, no. 1 (2010): 17-31; Gijs Mom, Colin Divall, and Peter Lyth, "Towards a Paradigm Shift? A Decade of Transport and Mobility History," in *Mobility in History: The State of the Art in the History of Transport, Traffic and Mobility*, ed. Gijs Mom, Gordon Pirie, and Laurent Tissot (Neuchâtel: Editions Alphil, 2009), 13-40; Mimi Sheller and John Urry, "The New Mobilities Paradigm," *Environment and Planning A* 38, no. 2 (2006): 207-226.

Bek, Patrick, *No Bicycle, No Bus, No Job: The Making of Workers' Mobility in the Netherlands, 1920-1990*. Taylor & Francis Group 2022
DOI: 10.5117/9789463723183_INTRO

ride it to work, he said. And, unlike his previous firm, the new employer did not offer company bus transportation. Because K. felt he had no viable mobility options to commute 11 km, he declined the job offer. The government agency, following the letter of the law, stopped K.'s unemployment benefit. Later, the Board of Appeal (Raad van Beroep) and Labor Council (Centrale Raad voor Arbeid) acknowledged that a daily cycle or moped commute came with "a certain inconvenience," especially in bad weather. Still, the Council deemed K. able-bodied enough to cycle to work. He should repair his bicycle or buy another one—second hand if necessary. They decided that "a healthy young man" bore responsibility for his own mobility access to work—a ruling that established key jurisprudence for future court cases.[2]

This legal landmark case illustrates the close relationship between (im)mobility and job accessibility. The controversy also shows different interpretations of who was responsible for the commute—and that the issue of how to get to work had become political. K.'s appeal and subsequent rejection both mark a pivotal moment in how the state and employers thought about who was responsible for facilitating workers' commute. Covering five industrial regions in the Netherlands since the 1920s, *No Bicycle, No Bus, No Job* shows how the locus of control shifted between workers, employers, and the government in addressing workers' (im)mobility. Workers and employers—against the backdrop of twentieth century economic booms and busts, wartime destruction and postwar recovery, periods of scarcity and affluence—were key in shaping the everyday commute. Until the 1970s at least, the state took a back seat. The global wave of deindustrialization and onset of neoliberal public governance, however, heralded a transformation. It left workers like K. to their own devices.

The problem has not ceased. Since 2003, research by Susan Kenyon and others on transport-related social exclusion provides ample evidence of how mobility barriers, (job) accessibility, and social exclusion reinforce each other into a downward spiral to poverty.[3] Following earlier international research, Dutch social geographers Jeroen Bastiaanssen, Karel Martens, and Gert Jan Polhuijs conclude in their 2013 case study of low-income jobseekers in the

2 Nationaal Archief, Archive no. 2.15.62, Directoraat-Generaal voor de Arbeidsvoorziening van het Ministerie van Sociale Zaken en Werkgelegenheid, 1970-1990, Inventory no. 2199, "Handleiding Passende Arbeid" (1977), C.R.v.B., 27 March 1975, R.S.V. 1975 no. 341; "Fietsen naar het werk," *Het Parool*, 1 October 1975, 23.

3 Susan Kenyon, Glenn Lyons, and Jack Rafferty, "Transport and Social Exclusion: Investigating the Possibility of Inclusion through Virtual Mobility," *Journal of Transport Geography* 10, no. 3 (2003): 207-219; Karen Lucas, "Transport and Social Exclusion: Where Are We Now?," *Transport Policy* 20 (2012): 105-113.

Netherlands' second largest city, Rotterdam, that being "without a driving license, without a bicycle, without public transit access" means "no job."[4] They have put the issue on the political agenda as a public responsibility: the state should get involved. Scholars have identified national and local government as key to breaking this vicious circle: ensuring mobility systems work for vulnerable social groups through regulations, safe roads, bicycle paths, and public transit subsidies.[5] While scholars routinely call for the state to play its part, they have so far overlooked the historical, but changing role of employers.

Employers are important actors in facilitating and shaping workers' everyday mobility. They lobby for infrastructures like public transit, roads, and bicycle paths for their workers. Perhaps surprisingly, in interwar America, Detroit Ford Motor Company supported a rapid-transit system to enable nearly a hundred thousand workers to access its sprawling River Rouge plant—and, as American scholars have detailed, shortly after the Second World War, switched to facilitating automobility and expressways so that workers could reach faraway industrial sites.[6] Employers also provided employees travel allowances per kilometer, individual travel budgets, lease plans for cars, and fiscal benefits when purchasing a bicycle.[7] Company bus transportation represents another, more direct intervention in lowering mobility barriers for car-less workers in remote (gateway) locations like ports and business parks near highways. Especially larger companies with sufficient financial means and political leverage can enhance job accessibility.[8]

4 Jeroen Bastiaanssen, Karel Martens, and Gert-Jan Polhuijs, "'Geen rijbewijs, geen fiets, geen ov-aansluiting, geen baan': Vervoersarmoede in Rotterdam-Zuid," *Verkeerskunde* 5 (2013): 44-50.

5 Jeroen Bastiaanssen, Daniel Johnson, and Karen Lucas, "Does Transport Help People to Gain Employment? A Systemic Review and Meta-Analysis of the Emperical Evidence," *Transport Reviews* 40, no. 5 (2020): 607-628.

6 In the Netherlands, Berkers and Oldenziel (2017) note that Dutch synthetic fiber industry AKU lobbied for bicycle paths to cater to the masses of cycling commuters to their Arnhem sites. Eric Berkers and Ruth Oldenziel, *Cycling Cities: The Arnhem and Nijmegen Experience* (Eindhoven: Foundation for the History of Technology, 2017), 26; Charles Hyde, "Planning a Transportation System for Metropolitan Detroit in the Age of the Automobile: The Triumph of the Expressway," *Michigan Historical Review* 32, no. 1 (2006): 59-95.

7 Verkeersonderneming, "Werkgeversaanpak," https://www.verkeersonderneming.nl/r10/, accessed May 7, 2021; Ministerie van Infrastructuur en Waterstaat, "Werkgevers stimuleren fietsgebruik medewerkers," Rijksoverheid, https://www.rijksoverheid.nl/onderwerpen/fiets/werkgevers-stimuleren-fietsgebruik-medewerkers, accessed June 18, 2021.

8 Verkeersonderneming, "Werkgeversaanpak"; Thomas Vanoutrive, "Commuting, Spatial Mismatch, and Transport Demand Management: The Case of Gateways," *Case Studies on Transport Policy* 7, no. 2 (2019): 489-496.

Business involvement in workers' mobility also comes with risks. According to the 2020 Dutch government report, *No Second-Class Citizens* (*Geen Tweederangsburgers*), powerful employers and employment agencies provide not just work for thousands of immigrants in agricultural, meat, and distribution industries, but also housing, health insurance, and transport. Several employers house migrant workers in cheap accommodation—for example vacant holiday parks far from work locations—and transport them by shuttle bus to worksites. They deduct substantial travel costs from workers' wages without providing compensation for long travel times or allowing workers the opportunity to live closer by, opt for mobility alternatives, and report exploitation. Employers are thus potentially key actors in reducing mobility barriers for workers when labor is scarce—though these cases also signal the mobility injustices that might arise when profits and control over workers rather than their well-being and social justice are guiding principles.[9] In other words, employers' involvement in facilitating workers' mobility to find and keep a job is precarious.

The reality stands in contrast to today's mobility discourse and practice. Upper and middle-class people tend to be highly mobile. Yet, low-income workers and jobseekers commute shorter distances because of what scholars call "limited travel horizons", experience severe cost and availability barriers, and rely more often on slower modes of transport.[10] They are forced to use failing public transit services—poorly connected to job locations and adjusted to working hours. They often must endure longer travel times and (socially) unsafe mobility. And they cannot access jobs in car-only areas or are forced to purchase a car, further straining already tight household budgets. Not just in car-oriented America. Case studies across the globe indicate this is a universal problem with local variations of mobility systems and power relations.[11] Even in the Netherlands, internationally renowned

9 Emile Roemer, *Geen Tweederangsburgers: Aanbevelingen om misstanden bij arbeidsmigranten in Nederland tegen te gaan* (The Hague: Aanjaagteam Bescherming Arbeidsmigranten, Rijksoverheid, 2020), 13-14, 20, 67.

10 Morris observes that low-income persons tend to have more limited travel horizons compared to middle- and upper-class people, largely as the result of being without a car or other options. Middle-class car drivers have a wider action radius and consequently more opportunity to land better paid jobs further afield. Kate Morris, "Research into travel horizons and its subsequent influence on accessibility planning and demand responsive transport strategies in Greater Manchester," paper presented at the *European Transport Conference*, Strasbourg, France (2006).

11 Bastiaanssen, Johnson, and Lucas, "Does Transport Help?," 607-628; Hans Jeekel and Karel Martens, "Equity in Transport: Learning from the Policy Domains of Housing, Health Care and Education," *European Transport Research Review* 9, no. 4 (2017): 1-13; Tobias Kuttler and Massimo Moraglio, *Re-Thinking Mobility Poverty: Understanding Users' Geographies, Backgrounds and Aptitudes* (London/New York: Routledge, 2020); Karen Lucas, "The Role of Transport in the Social

for its strongly embedded bicycle and car regime, plus well-developed albeit expensive public transit, precarious mobility is a daily reality for many.[12]

These predicaments are not new, even though some scholars suggest otherwise. According to Tobias Kuttler and Massimo Moraglio, the theme of transport-related social exclusion first appeared in academic debates through geographer Karen Lucas around 2004.[13] For the Netherlands, Karel Martens, Marnix ten Holder, and Jurriën Thijssen presented their work in 2011 as one of the first to address the problem in Dutch academic circles and policy debates.[14] These claims illustrate the systemic lack of historical awareness. The phenomenon has a much longer tradition. Sociologist Colin Pooley signaled a widening inequality gap in British people's accessibility around 1970, when car-centered transport and land-use planning raised mobility barriers for people without a car.[15] As I will show—something that no scholar has yet addressed—similar issues were also raised in the Netherlands half a century ago.

Indeed, today's predicaments are not the sole effect of recent political decisions. They have a history. Mobility systems have long lifespans—and reflect decisions made in the past. In the words of Frank Schipper, Martin Emanuel, and Ruth Oldenziel, "it takes decades to build—and by the same token to unbuild—systems that include infrastructures (from bridges to airports), as well as the institutions (from semi-governmental transport agencies to powerful lobbies) sustaining them." Transforming mobility systems entails

Exclusion of Low Income Populations in South Africa," paper presented at the *World Conference on Transport Research Society*, Lisbon, Portugal, July 11-15, 2010, https://www.wctrs-society.com/wp-content/uploads/abstracts/lisbon/general/01390.pdf, 1-21, accessed October 31, 2021; Deborah Salon and Sumila Gulyani, "Mobility, Poverty, and Gender: Travel 'Choices' of Slum Residents in Nairobi, Kenya," *Transport Reviews* 30, no. 5 (2010): 641-657; Thomas Vanoutrive, "Minder inkomen, dus minder mobiel, dus minder kansen," *Armoede en Sociale Uitsluiting: Jaarboek 2018*, edited by Jill Coene et al. (Leuven/The Hague: Acco, 2018), 277-290; Dominic Villeneuve and Vincent Kaufmann, "Exploring the Causes of Social Exclusion Related to Mobility for Non-Motorized Households," *Transportation Research Board* 2674, no. 8 (2020): 911-920.

12 Bastiaanssen, Martens, and Polhuijs, "'Geen rijbewijs, geen fiets, geen ov-aansluiting, geen baan'", 44-50; Rob van der Bijl and Hugo van der Steenhoven, *Gesprekken over gebrekkige mobiliteit: Vervoersarmoede in de grote stad ontrafelen* (Amsterdam/Utrecht: Favas/HugoCycling, 2019), 7-8, 12-16; Peter Jorritsma et al., *Mobiliteitsarmoede: Vaag begrip of concreet probleem?* (The Hague: Kennisinstituut voor Mobiliteitsbeleid, 2018), 27-28.

13 Kuttler and Moraglio, *Re-Thinking Mobility Poverty*, 6-7; Karen Lucas, *Running On Empty: Transport, Social Exclusion and Environmental Justice* (Bristol: The Policy Press, 2004).

14 Karel Martens, Marnix ten Holder, and Jurriën Thijssen, "Vervoersarmoede Bestaat," *Verkeerskunde* 2, no. 11 (2011): 34-38.

15 Colin Pooley, "Mobility, Transport and Social Inclusion: Lessons from History," *Social Inclusion* 4, no. 3 (2016): 100-109.

understanding the long-term development paths and "supporting coalitions of vested interests built around them over decades."[16] Today's academic and policy debates on mobility poverty often ignore this historical perspective. Adopting a long-term approach is key to helping decisionmakers analyze causes, identify alternative mobility futures, and discern the power relations that drive historical trends. Without such a perspective, academic and policy debates run the risk of being short-terminist, piecemeal, and eventually having limited or even adverse effects.[17] In taking up this call, *No Bicycle, No Bus, No Job* reveals that workers' (im)mobility has been the outcome of social processes driven by workers, employers, and the government.

How Workers' Travel was Controlled in Many Ways

Labor historians have a long tradition of analyzing power and agency in the context of working-class life, labor relations, and the organization of work. They have shown how freedom and unfreedom, autonomy and heteronomy, are often different sides of the same coin. In his seminal work *The Making of the English Working Class* (1963), labor historian E.P. Thompson criticized the reductionist tendency to describe the working class as an amorphous unchangeable entity or uneducated mass that acted on impulse and emotion. Such writing obscured workers' experiences, aspirations, moral convictions, and ingenuity. Workers were not merely the victims of capitalist history. "The working class made itself as much as it was made" through solidarity, collectivism, and political action, Thompson famously wrote.[18] Building on his work, scholars have shown how important unions have been in representing skilled and unskilled workers to leverage power collectively and create agency in shaping their lives. Synthesizing many international studies, Marcel van der Linden explains unions came in many shapes, but essentially enabled collective bargaining over employees' rights, wages, and working conditions with strikes as ultimate political levers.[19]

16 Frank Schipper, Martin Emanuel, and Ruth Oldenziel, "Introduction: Historicizing Sustainable Urban Mobility," in *A U-Turn to the Future: Sustainable Urban Mobility since 1850*, edited by Martin Emanuel, Frank Schipper, and Ruth Oldenziel (New York: Berghahn Books, 2020): 1-26, here 3-4.

17 Peter Norton, "Urban Transport and Mobility in *Technology and Culture*," *Technology and Culture* 61, no. 4 (2020): 1197-1211, here 1205; Erik van der Vleuten, "History and Technology in an Age of 'Grand Challenges': Raising Questions," *Technology and Culture* 61, no. 1 (2020): 260-271.

18 Edward Palmer Thomspon, *The Making of the English Working Class* (London: Penguin, 2013 [1963]), 213.

19 Marcel van der Linden, *Workers of the World: Essays Toward a Global Labor History* (Leiden/Boston: Brill, 2011), 219, 225

While recognizing workers' agency, labor historians also pointed out its limits, starting with Thompson who explained how since the Industrial Revolution, state and factory managers imposed synchronic forms of time and work discipline on working people, curtailing their freedom of choice.[20] The struggle over employers' control and workers' agency has been a key theme in labor history. Other critical thinkers have also brought under scrutiny the limits of individual freedom in modern capitalist society. In a society organized around mentalities of efficiency, rationality, and social control, to what extent could people still autonomously decide the direction of their own lives?[21] While my work brings to the fore workers' agency, it also shows that changing and often uneven power relations profoundly shaped their ability to decide whether and how to travel.

For theorizing how different modes of power operate in modern and liberal societies, French philosopher and historian Michel Foucault has been highly influential. Foucault argued that modern forms of power—what he referred to as "governmentality"—seek "to incite, reinforce, control, monitor, optimize, and organize the forces under it." Governmentality involves "a power bent on generating forces, making them grow, and ordering them, rather than one dedicated to impeding them, making them submit, or destroying them."[22] During the Industrial Revolution, new modes of power emerged, not just for organizing work and capital. They centered on "the body as a machine, optimizing its capabilities, increasing its usefulness and docility, integrating it into systems of efficient and economic controls."[23] It enabled those in power—like factory managers—to meticulously control the physical movements of their workers (whom he called "bodies") and impose on them what Foucault defined as a "relationship of docility-utility." This control was exercised through the physical arrangement of built environments, work schedules, and the manipulation of machines and factories.[24]

Foucault's work has greatly influenced historical analyses of power. Foucault helps to conceptualize how state and company power operates, and how power

20 Edward Palmer Thompson, "Time, Work-Discipline, and Industrial Capitalism," *Past & Present*, no. 38 (1967): 56-97.

21 See among other works: Ivan Illich, *Deschooling Society* (San Francisco: Harper & Row, 1971), xix-xx; *Tools for Conviviality* (San Francisco: Harper & Row, 1973), 50-53; Herbert Marcuse, *One-Dimensional Man* (London/ New York: Routledge, 2002 [1964]), 3-20; Max Weber, *Economy and Society: An Outline of Interpretive Sociology* (Berkeley: University of California Press, 1978), lix.

22 Michel Foucault, *Power/Knowledge: Selected Interviews and Other Writings, 1972-1977* (New York: Pantheon Books, 1980), 136, 139, 141-142, 144.

23 Michel Foucault, *History of Sexuality*, vol. 1 (London: Penguin Books, 1978), 139, 144.

24 Michel Foucault, *Discipline and Punish: The Birth of the Prison* (London: Penguin/Random House, 2020 [1977]), 136-138.

relations between workers shift. A common criticism of his work, however, is that he regarded bottom-up attempts to resist dominant power structures pointless and nullified individual agency. Even in cases where individuals could be regarded as self-governing, Foucault deemed their actions to be curtailed by power relations and disciplinary mechanisms.[25] I explore this tension in the context of workers' mobility, tracing who controlled the everyday commute. Thus, we should not see the workers on these pages as docile subjects at the mercy of a dominant state or factory manager. Workers often acted as autonomous, self-governing agents too. They did so in a changing playing field of power relations that shifted from paternalism to neo-liberalism.

Labor historians have meticulously researched how such control over workers came about. They have unraveled industrial-capitalist politics of control over workers in terms of time, space, and movement. In *Handbook Global History of Work*, Karin Hofmeester and Marcel van der Linden synthesize an extensive body of scholarship to explain that modern labor management started back in the mid-eighteenth century with the development of industrial capitalism. Factory owners and later managers, eager to accumulate surplus value, reduced labor costs by imposing time-discipline, training, and scientific management.[26] Machines and factory floors were designed in such a way that managers could detect loitering workers and reduce any unnecessary actions that might hamper the workflow.[27] Applying formal and informal rules, employers stipulated when, where, and how labor had to be performed, by whom, and for what reward (or penalty). Later, this ideology of efficiency spread from the United States to Europe and beyond through magazines, books, trade shows, and consultancy firms.[28] These studies focus on what on what happened on work sites.

25 Kurt Borg, "Conducting Critique: Reconsidering Foucault's Engagement with the Question of the Subject," *Symposia Melitensia*, no. 11 (2015): 1-15, here 1-2, 14.

26 Marcel van der Linden, "Work Incentives and Forms of Supervision," in *Handbook Global History of Work*, ed. Karin Hofmeester and Marcel van der Linden (Berlin/Boston: De Gruyter Oldenourg, 2018), 469-470, 479-481.

27 Andrew Herod, "Social Engineering through Spatial Engineering: Company Towns and the Geographical Imagination," in *Company Towns in the Americas: Landscape, Power, and Working-Class Communities*, ed. Oliver Dinius and Angela Vergara (Athens, GA: University of Georgia Press, 2011), 30; Arwen Mohun, "Labor and Technology," in *A Companion to American Technology* ed. Carroll Pursell (Malden, MA: Blackwell, 2008), 214-215, 219-221; Erik van Vleuten, Ruth Oldenziel, and Mila Davids, *Engineering the Future, Understanding the Past: A Social History of Technology* (Amsterdam: Amsterdam University Press, 2017), 77; Frans van Waarden, "Organisatie, arbeid en ondernemersbeleid in de twentse katoenindustrie," in *Ter Elfder Ure* 33, edited by Hugues Boekraad et al. (Nijmegen: SUN, 1983), 157-200, here 158, 170-180.

28 Linden, "Work Incentives and Forms of Supervision," 469-470, 479-481; Vleuten, Oldenziel, and Davids, *Engineering the Future*, 77-78; Waarden, "Organisatie, arbeid en ondernemersbeleid

I show how managers also interfered with what assembly-line workers, dockers, miners, steel workers, and textile workers did outside the factory gates. Historians have traced how following American Fordism, industrialists worldwide believed that investments in workers' quality of life would increase well-being and productivity, thus reduce labor turnover and conflict in the form of week-long strikes. Historian Howard M. Gitelman notes that under the guise of "industrial welfare" (or "welfare capitalism"), managers attempted to make workers' bodies more productive by providing proper nutrition, housing, and medical care. In contrast to the more discretionary, nineteenth-century paternalism found in family-owned firms, newly established social affairs departments organized this form of paternalism and labor control more systematically.[29] Dutch historians have detailed these trends as well: electronics company Philips and steelworks company Hoogovens established their social services in the interwar years. Amid the postwar push for industrial growth, welfare programs found wider application. Company-owned guesthouses, neighborhoods, and towns were common practice across the industrializing world. For the Netherlands too, such initiatives symbolized companies' well-intended industrial paternalism. Driven by global economic forces and profit, these also symbolized their control over workers' lives.[30] Labor history has detailed these trends in industrial capitalism throughout the world. Still, company housing was an important form of control outside the factory gates, but limited in terms of numbers.

That is not the case for workers' mobility.[31] Scholars have ignored this aspect of workers' life as well as the forces that shaped that experience. A

in de twentse katoenindustrie," 158, 170-180.

29 Howard M. Gitelman, "Welfare Capitalism Reconsidered," *Labor History* 33, no. 1 (1992): 5-31.

30 Bram Bouwens et al., *Door staal gedreven: Van Hoogovens tot Tata Steel, 1918-2018* (Bussum: Uitgeverij Thoth, 2018), 45-47; Gerard Kuijs, *De vrees voor wat niet kwam: Nieuwe arbeidsverhoudingen in Nederland 1935-1945, aan het voorbeeld van de Twentse textielindustrie* (Amsterdam: Vossiuspers UvA, 2010), 176-179; Charles S. Maier, "Between Taylorism and Technocracy: European Ideologies and the Vision of Industrial Productivity in the 1920s," *Journal of Contemporary History* 5, no. 2 (1970): 27-61; Frank Pot, *Zeggenschap over beloningssystemen 1850-1987* (Leiden: NIPG/TNO, 1988), 172; Sjef Stoop, *De sociale fabriek: Sociale politiek bij Philips Eindhoven, Bayer Leverkusen en Hoogovens IJmuiden* (Utrecht: Stenfert Kroese, 1992), 21-22, 29-33, 44-46; Vleuten, Oldenziel, and Davids, *Engineering the Future*, 77-79; Jaap Vogel, *Nabije vreemden: Een eeuw wonen en samenleven*, Cultuur en Migratie in Nederland (The Hague: SDU Uitgevers, 2005).

31 Since the early 2000s, a mobilities turn in transport history widened research topics and approaches, shifting away from roads, vehicles, physical infrastructures, to people and things moving between places, sensitive to underlying politics, social meanings, and practices. Simone Fari and Massimo Moraglio, "Future Mobilities: A Challenge for Economic and Business Historians," in *43rd Annual Economic and Business History Society Conference* (University of Jyväskylä2018), 2, 7-8, 14; Gijs Mom, "What Kind of Transport History Did We Get? Half a Century

literature review of *Labor History* and *The International Review for Social History*—both leading journals in labor history—shows how daily commuting became a more common job market strategy for blue-collar workers since the late nineteenth century. These studies, however, do not detail how people travelled to work. Nor do they address the underlying power issues that shaped workers' mobility.[32] This omission is remarkable because, unlike company housing, controlling how employees got to work involved a larger portion of the workforce. Examining the phenomenon extends our understanding of how workers are governed.

Numerous specialists in mobility history have addressed workers' daily journeys as a phenomenon without going into detail or mention workers' experience only in passing. These mobility history studies imply workers—though not their research focus—had agency in choosing how they commuted. In late-nineteenth and early twentieth-century Europe, most people walked to work because their workplace was usually near where they lived. In Enschede—the epicenter of the Dutch textile industry—for instance, workers from the newly built working-class neighborhoods walked to work.[33] For longer distances, some commuted by tram and train. Colin

of JTH and the Future of the Field," *The Journal of Transport History* 24, no. 2 (2003): 121-138, here 122-123, 126, 128, 130-132; Mom, Divall, and Lyth, "Towards a Paradigm," 14, 17, 19, 21-23; Norton, "Urban Transport and Mobility in *Technology and Culture*," 1201, 1204.

32 *Labor History* and *The International Review for Social History*: Robert Baker, "Socialism in the Nord, 1880–1914. A Regional View of the French Socialist Movement," *The International Review of Social History* 12, no. 3 (1967): 357-389; Jan Breman, "Industrial Labour in Post-Colonial India II: Employment in the Informal-Sectory Economy," *The International Review of Social History* 44, no. 3 (1999): 451-483; Simon Constantine, "Migrant Labour in the German Countryside: Agency and Protest, 1890–1923," *Labor History* 47, no. 3 (2006): 319-341; Martin Daunton, "Miners Houses: South Wales and the Great Nortern Coalfield, 1880-1914," *The International Review of Social History* 25, no. 2 (1980): 143-175; Elena Dinubila, "Fight for Meaning: Representations and Work Experiences in a Greenfield Automotive Plant," *Labor History* 61, no. 1 (2020): 60-73; Alf Lüdtke and William Templer, "Polymorphous Synchrony: German Industrial Workers and the Politics of Everyday Life," *The International Review of Social History* 38, Supplement 1 (1993): 39-84; Ian Kerr, "On the Move: Circulating Labor in Pre-Colonial, Colonial, and Post-Colonial India," *The International Review of Social History* 51, Supplement 14 (2006): 85-109; Can Nacar, "'Our Lives Were Not as Valuable as an Animal': Workers in State-Run Industries in World-War-II Turkey," *The International Review of Social History* 54, Supplement (2009): 143-166; Dhiraj Nite, "Employee Benefits, Migration and Social Struggles: An Indian Coalfield, 1895-1970," *Labor History* 60, no. 4 (2019): 372-391; Randall Patton, "Textile Organizing in a Sunbelt South Community: Northwest Georgia's Carpet Industry in the Early 1960s," *Labor History* 39, no. 3 (1998): 291-309; Peter Scott, "Women, Other 'Fresh Workers', and the New Manufacturing Workforce of Interwar Britain," *The International Review of Social History* 45, no. 3 (2000): 449-474.

33 Adri Albert de la Bruhèze, "Enschede: An Experiment in Cycling," in *Cycling Cities: The European Experience*, edited by Ruth Oldenziel et al. (Eindhoven: Foundation for the History of Technology, 2016), 41-51, here 41.

Pooley and Jean Turnbull signaled major transformations in commuting patterns in twentieth century Britain, which appear broadly consistent across Europe.[34] Since the interwar period, three major shifts occurred in workers' mobility. First, workers discovered bicycles in the interwar period. Historians Adri Albert de la Bruhèze and Frank Veraart show in *Fietsverkeer in praktijk en beleid in de twintigste eeuw* (1999) that across interwar Europe, cycling was booming among the working-classes. Later *Cycling Cities: The European Experience* (2016), edited by Ruth Oldenziel, Martin Emanuel, Adri Albert de la Bruhèze, and Frank Veraart expanded on this insight.[35] Second, workers found affordable alternatives to fixed-route rail transport in paratransit taxi and bus services. In early twentieth century France, manual workers could commute via collective taxis for two francs, a fifth of first-class rail fares. In the Netherlands, similar bus services mushroomed, as Gijs Mom and Ruud Filarski show in *Van Transport naar Mobiliteit,* though they do not highlight their significance for working-class commuters.[36] Third, in the second half of the twentieth century, a growing number of working-class households purchased mopeds and cars. Car ownership took off later in the Netherlands compared to neighboring Belgium, Britain, and Germany, before catching up fast, spurred by the postwar economic boom and the government's liberalization of wages in 1963. According to historians Gijs Mom, Johan Schot, and Peter-Eloy Staal, by 1970, cars had also become a more common option for commuting in the Netherlands.[37] Matters of agency and power remain largely unaddressed in the mobility scholarship.

34 Colin Pooley and Jean Turnbull, "The Journey to Work: A Century of Change," *Area* 31, no. 3 (1999): 281-292; "Modal Choice and Modal Change: The Journey to Work in Britain Since 1890," *Journal of Transport Geography* 8, no. 1 (2000): 11-24.

35 Adri Albert de la Bruhèze and Frank Veraart, *Fietsverkeer in praktijk en beleid in de twintigste eeuw* (The Hague: Foundation for the History of Technology, 1999); Ruth Oldenziel et al., *Cycling Cities: The European Experience* (Eindhoven: Foundation for the History of Technology, 2016).

36 Ruud Filarski, "De coördinatiecrisis," in *Van transport naar mobiliteit: De mobiliteitsexplosie (1895-2005)*, ed. Gijs Mom and Ruud Filarski (Zutphen: Walburgpers, 2008), 203-235; Mathieu Flonneau, "Collective Taxis in 1930s Paris: A Contribution to an Archeology of 'Uberization,'" *The Journal of Transport History* 39, no. 1 (2018): 12-24; Gijs Mom, "Clashes of Cultures: Road vs. Rail in the North Atlantic World during the Interwar Coordination Crisis," in *The Organization of Transport: A History of Users, Industry, and Public History*, ed. Massimo Moraglio and Christopher Kopper (New York: Routledge, 2015), 18-37.

37 Doreen Ewalds, Ger Moritz, and Michel Sijstermans, *Bromfietsen in Nederland* (The Hague/Heerlen: Centraal Bureau voor Statistiek, 2013), 4; Gijs Mom, Johan Schot, and Peter-Eloy Staal, "Werken aan mobiliteit: de inburgering van de auto," in *Techniek in Nederland in de twintigste eeuw: Transport, communicatie*, edited by Harry Lintsen and Johan Schot (Zutphen: Walburgpers, 2002): 45-73, here 70; Peter-Eloy Staal, *Automobilisme in Nederland: Een geschiedenis van gebruik, misbruik en nut* (Zutphen: Walburgpers, 2003), 118.

Less well described is how mobility is produced, under what conditions, and who has a say in decision-making. This also affects whether and how workers get to work. In *Mobility Justice*, sociologist and historian Mimi Sheller argues that whether, when, and how people move are political and moral questions. Mobility is never politically neutral, but "full of frictions, viscosity, stoppages, and power relations. ... always contingent, contested, and performative. ... never free but ... in various ways always channeled, tracked, controlled, governed, under surveillance and unequal—striated by gender, race, ethnicity, class, caste, color, nationality, age, sexuality, disability, etc., which are all in fact experienced as effects of uneven mobilities."[38] A mobility justice perspective not only focuses on the actual movement from A to B and how available mobility options are distributed over society. It also involves understanding how power relations and systems of governance enable or prohibit movement.

Commuting—the ability to get to work in the first place—is thus not a simple matter of choosing whether, how and when to travel. Power and privilege are what determine one's field of action. Compared to privileged highly mobile people, low-income workers and jobseekers have fewer mobility options. This is not only down to greater physical distances from jobs or physical abilities to travel, but also levels of network capital, according to mobility scholars John Urry and Mimi Sheller. Marginalized people's capacity to engender and maintain access to financial capital and social networks for economic and practical benefit is generally more limited. Typically, they cannot choose freely between a wide range of affordable mobility alternatives, nor do they have a say in political decisions that affect how they get around. Consequently, this reduces their ability to overcome mobility barriers and curtails their autonomy to decide how and when to travel.[39]

A few mobility historians have detailed how the state played a key role in shaping the preconditions that enabled workers to commute. Social reform concerns about workers' living conditions were an important incentive for developing rail transportation in the late 1800s. As more unskilled workers gained employment in urban docks, factories, and construction sites, government and employers' concerns about hygiene, impoverishment, and proletarization fueled the planning of new working-class housing

38 Mimi Sheller, *Mobility Justice: The Politics of Movement in an Age of Extremes* (London/New York: Verso, 2018), XV, 10-11.

39 Tobias Kuttler and Massimo Moraglio, *Re-Thinking Mobility Poverty: Understanding Users' Geographies, Backgrounds and Aptitudes* (London/New York: Routledge, 2020), 10; Mimi Sheller, "Theorising Mobility Justice," *Tempo Social* 30, no. 2 (2018): 26; *Mobility Justice*, 159-160; John Urry, *Mobilities* (Cambridge/Malden: Polity, 2007), 197.

outside urban centers (garden cities). As travel distances between home and work increased across Western Europe, the demand for rail connections rose, mobility historians Colin Divall and Winstan Bond detail in *Suburbanizing the Masses*.[40] According to Bond, paternalistic ideas about improved working-class well-being and housing "away from the smoke, disease and congestion," went hand in hand with early twentieth century state-subsidized rail expansion connecting working-class neighborhoods with industrial sites.[41] Special workmen fares also enabled workers from far and wide to access industrial jobs across Western Europe.[42] Greet De Block argues that in Flanders, these developments effectuated "the 'emancipation of the working class' in an economically liberal meaning, providing material equality to laborers that were isolated in rural areas without being able to go out according to their interests and needs."[43] This also applied to the Netherlands. Even though rail-based travel, like elsewhere, declined sharply in later decades, the Dutch government financially supported its railways, maintaining a basic infrastructure.[44]

Road and bicycle path construction was another government task, as mobility historians show. In the Netherlands, like in other European countries, road construction took off in the interwar period. Politicians, policymakers, and engineers governed this process. Initially, national

40 Elisabeth Bervoets, "Modernisering van de woningbouw 1890-1970: een bijzonder patroon van technische vernieuwing," in *Techniek in Nederland in de twintigste eeuw. Deel 6. Stad, bouw, industriële productie*, ed. Johan Schot, et al. (Eindhoven/Zutphen: Foundation for the History of Technology /Walburgpers, 2003): 110-117, here 119; Hans Buiter, "Werken aan sanitaire en bereikbare steden, 1880-1914," in *Techniek in Nederland in de twintigste eeuw. Deel 6: Stad, bouw, industriële productie*, ed. Johan Schot, et al. (Eindhoven/Zutphen: Foundation for the History of Technology /Walburgpers, 2003): 25-49, here 41; Colin Divall and Barbara Schmucki, "Introduction," in *Suburbanizing the Masses: Public Transport and Urban Development in Historical Perspective* ed. Colin Divall and Winstan Bond (London/New York: Routledge, 2003), 10-11.

41 Winstan Bond, "The Flawed Economics and Morality of the American Uniform Five-cent Fare," in *Suburbanizing the Masses*: 49-78, here 65.

42 Paolo Capuzzo, "Transport and (sub)urban development: Between politics and technology: transport as a factor of mass suburbanization in Europe, 1890-1939," in *Suburbanizing the* Masses: 23-48, here 25-27, 30-31, 35; Tomas Ekman, "Vision in Solid Form: A Comparison Between Two Solutions to the Traffic Probelm in Stockholm, 1941 and 1992," in *Suburbanizing the Masses*, 181; Dieter Schott, "Suburbanizing the Masses for Profit or Welfare: Conflict and Cooperation Between Private and Municipal Interests in German Cities, 1890-1914," in *Suburbanizing the Masses*, 79-80.

43 Greet De Block, "Urbanizing the Countryside: Rails, Workers and Commuting in South-West Flanders, Belgium, 1830-1930," in *Cultural Histories of Sociabilities, Spaces and Mobilities*, ed. Colin Divall (London: Routledge, 2015): 53-66, here 53-55, 58.

44 Gijs Mom and Ruud Filarski, *Van Transport naar Mobiliteit: De Mobiliteitsexplosie 1895-2005* (Zutphen: Walburgpers, 2008), 390-393.

funding was allocated for constructing interlocal highways and Dutch city planners and policymakers increasingly redeveloped streets for projected car drivers, even though pedestrians and cyclists were still the major road users. This process accelerated in the postwar decades, encouraged by U.S. Marshall Aid and the technocratic ambitions of car-centered modernity. Historian Henk-Jan Dekker shows in *Cycling Pathways* that bicycle path construction also took off in interwar Netherlands. Building on the work of Veraart, Albert de la Bruhèze, and Oldenziel et al., Dekker shows that Dutch decision-makers envisioned bicycle paths would help get cyclists out of the way of car drivers, but also supported commuter cycling. In the postwar decades, until the bicycle revival in the 1970s, bicycle path construction was pushed to the sidelines in car-centered transportation and land-use planning. Still, it remained an important provincial and city government task.[45] Roads and bicycle paths formed the basic infrastructures that potentially enabled workers' movements, funded and planned by the state. Scholars have focused less on whether workers could and did actually use such infrastructures, though historians provide ample evidence that engineers and local authorities took these issues seriously.

A few mobility scholars show that, unlike its commitment to enhance workers' mobility by rail and road, the government sought to control workers' movements on foot, bicycle, and bus—popular among the working-classes. Bicycles and buses were subjected to intensive regulations by local authorities and the national government. These efforts aimed to control rather than enhance mobility. The severe interventions reveal how state authorities curtailed certain movements in the name of social order, modernity, and fair competition. When automobility emerged, car boosters and their allies restricted walking and cycling practices through direct control, regulation, and traffic educational campaigns. Liability for traffic accidents increasingly shifted from car drivers to pedestrians and cyclists' allegedly unruly bodies and undisciplined minds. Working-class cyclists were policed and schooled until they fitted this powerful and normative discourse. In many instances they were excluded from using certain spaces, historians Oldenziel and Albert de la Bruhèze claim in "Contested Spaces: Bicycle Lanes in Urban

45 Henk-Jan Dekker, *Cycling Pathways: The Politics and Governance of Dutch Cycling Infrastructure, 1920-2020* (Amsterdam: AUP, 2021), 84-85, 106, 118-119, 205-207; Gijs Mom, "Roads Without Rails: European Highway Network-Building and the Desire for Long-Range Motorized Mobility," *Technology and Culture* 46, no. 4 (2005): 745-772, here 747; Mom and Filarski, *Van Transport naar Mobiliteit (II)*, 200-201, 311-325.

Europe."[46] Similarly, paratransit bus operators were subjected to intensifying state regulation, motivated by the state's vested interests in rail transport. Mom argues these interventions were rooted in the interwar desire for order, fueled by middle-class fears of political revolution and European decision-makers seeking "harmony and spatial balance" Technocratic policymakers and transport economists envisioned that they could "win the wilderness over to order," in this case by prohibiting bus operators from freely adjusting fares, timetables, and routing—the very aspects that made bus transport popular with the public. Such attempts to govern mobility practices ultimately created a web of control over people's movements, according to Mom.[47]

Thus far, historians have not focused on the employers' role, highlighting that governments were the key agents in mobility development as part of a systematic state planning effort. Mom and Filarski point out that road construction was a negotiated process, guided by the interests of various non-state actors like tourist organization ANWB and road-building associations. Dekker details that the initiators in bicycle path construction were cycling citizens and the ANWB.[48] In both labor and mobility histories, the role of workers and employers, however, remains underexposed.

Mobility historians sporadically mention employers' lobbying efforts in railway, road, and bicycle path construction.[49] For the postwar era, labor and business historians only note in passing company interventions in mobilizing cheap rural labor. Erik Nijhof notes that in the decades following the Second World War, company-organized buses were essential for Rotterdam's port companies to attract cheap labor from rural regions.[50] Jan Zwemer's postwar history on the province of Zeeland confirms that buses enabled jobseekers in more remote areas to escape rural poverty by getting access to distant, better paid industrial jobs.[51] Serge Langeweg makes a similar observation for

46 Jennifer Bonham and Peter Cox, "The Disruptive Traveller? A Foucauldian Analysis of Cycleways," *Road and Transport Research* 19, no. 2 (2010): 42-53, here 15, 18-19; Ruth Oldenziel and Adri Albert de la Bruhèze, "Contested Spaces: Bicycle Lanes in Urban Europe, 1900-1995," *Transfers* 1, no. 2 (2011): 29-49.

47 Mom, "Clashes of Cultures," 18-21, 28.

48 Dekker, *Cycling Pathways*, 171-180; Mom and Filarski, *Van Transport naar Mobiliteit (II)*, 390-393.

49 Berkers and Oldenziel, *Cycling Cities*, 26; Dekker, *Cycling Pathways*, 84-85, 106, 118-119, 205-207; Hyde, "Planning a Transportation System for Metropolitan Detroit in the Age of the Automobile," 59-95; Mom and Filarski, *Van Transport naar Mobiliteit (II)*, 200-201, 311-325.

50 Erik Nijhof, *'Gezien de dreigende onrust in de haven': De ontwikkeling van de arbeidsverhoudingen in de Rotterdamse haven 1945-1965* (Amsterdam: IISG, 1988), 138-142.

51 Jan Zwemer, "De pendelarbeiders van Tholen en Sint-Philipsland," in *Zeeland 1950-1965*, ed. Jan Zwemer (Vlissingen: Uitgever Den Boer/De Ruiter, 2005): 401-428.

the Southeast Limburg mining region, in *Mijnbouw en Arbeidsmarkt*.[52] Bram Bouwens' business history of Dutch steel company Hoogovens mentions that instead of housing employees near blast furnace sites in IJmuiden, the postwar period marked a shift to mobilizing workers from the wider region by bus.[53] Despite such observations, these historical works do not delve deeper into the company governance of workers' commute. Moreover, they reduce workers to mere bus passengers, instead of seeing them as shaping agents who faced dilemma's in selecting mobility options in the first place to get to work.

Industrial capitalism could not have thrived without the appropriate social and material preconditions that enabled the accumulation of surplus value. Government and employers' ability to configure these preconditions has been an essential factor in making the production system work.[54] Company control over workers' movements played a central role in this endeavor. This not only applied to the shopfloor and company housing. It also shaped how workers commuted until deindustrialization, when employers' role diminished.

Mobility in Key Dutch Industrial Centers

No Bicycle, No Bus, No Job traces how the state, industrial employers, and workers shaped twentieth century commuting in Dutch industrial centers. I do not discuss commuting as a phenomenon, but aim to reveal what factors enabled or inhibited the mobility of workers with a lower socioeconomic status. The book covers most of the twentieth century, specifically the emergence and decline of manufacturing industries in the Netherlands between 1920 and 1990. It travels along with workers through the interwar Great Depression, war and postwar age of destruction and scarcity, then postwar reconstruction and subsequent economic growth (1945-1973), ending in the recession after the 1973 oil crisis, global wave of deindustrialization, and onset of neoliberal public governance. This periodization makes it possible to unravel the relationship between workers' mobility and industrial-capitalist company governmentality.

52 Serge Langeweg, *Mijnbouw en arbeidsmarkt in Nederlands-Limburg: Herkomst, werving, mobiliteit en binding van mijnwerkers tussen 1900 en 1965* (Hilversum: Verloren, 2011), 63-72.

53 Bouwens et al., *Door staal gedreven*, 119-120.

54 Herod, "Social Engineering," 26, 37.

Figure 1: The locations of the five case studies in the Netherlands

The book details how in mobilizing local, regional, and foreign labor forces, Dutch industrial employers used their power to develop mobility policies and infrastructures. The case studies are representative for the country and the era. I focus on major employers in five key industrial centers in the Netherlands (fig. 1), trailblazers for other (smaller) industries: Eindhoven's Philips electronics factories, IJmuiden's steel company Hoogovens, Limburg's mines, Twente's textile manufacturing, and Rotterdam's and Schiedam's docks.

Limburg's mines were located in the southern-most tip of the Netherlands, a hilly region bordering Belgium and Germany. Before mining expansion, the region's farmers earned a living on the land or in small manufacturing industries. Further north, the new town Eindhoven—a merger of villages like Woensel and Strijp—became home to national industries like car manufacturer DAF and electronics firm Philips. Near the eastern border with Germany, in the dispersed urban region of Twente, textile mills and machine factories were a common sight. Moving west to the North Sea coast, one finds IJmuiden, amid scattered towns and villages where people traditionally lived off agriculture and fishing before the steel plant Hoogovens opened in 1923. Further south, port city Rotterdam—together with bordering Schiedam—was a booming economic center with firms like Van Nelle tobacco factory, transshipment companies, and shipyards. Over centuries, Rotterdam had developed into a polycentric urban region, close to other major Dutch cities like The Hague, in what today is known as the Randstad conurbation. During the twentieth century, Rotterdam's port district expanded westward towards the North Sea coast, away from major residential areas.

The geographically diverse industries encompassed the large companies—not small or medium-sized enterprises—that were key for the Netherlands' twentieth century industrialization. These selected companies provided jobs for a diverse and large-scale workforce: hundreds of thousands of unskilled and semi-skilled men and women from near and far who dug for coal, washed cotton, operated quay-side cranes, stapled steel sheets, and wired electronic devices among many other tasks. The corporations hired both Dutch and foreign workers in the postwar era; they included male-dominated and unionized industries like mining and shipping as well as electronics and textiles companies that provided employment to many young, unmarried, and rural women; and male migrant and ununionized workers from impoverished rural areas in Belgium, Germany, Spain, Turkey, Morocco seeking for better opportunities and political refugees from the former Dutch colony. The representative case studies allow me to compare and contrast historical trends in workers' mobility and identify potential social differentiation in workers' commute at the intersections of class, gender, ethnicity, and geographic location.

Until the 1960s, most low-skilled workers in the Netherlands were in heavy manufacturing industries. Like elsewhere in Europe, Dutch industrial employers recruited workforces based on relative cheapness, flexibility, and often weak unionization, not seldom from (rural) regions with rising unemployment. In the interwar period, managers often replenished urban labor forces with workers from the surrounding countryside and across the border. In the postwar push for industrialization, Dutch managers attracted more unskilled workers, who would accept dirty work, unpaid over-hours, lower wages, and complained less about poor working conditions. These were often young men and women, rural jobseekers, and migrant workers, characterized by low unionization and the least power to demand higher wages. Current research identifies these groups as the ones most often facing mobility barriers and with few resources to overcome mobility poverty.[55]

55 Bastiaanssen, Johnson, and Lucas, "Does Transport Help," 607-628; Kenyon, Lyons, and Rafferty, "Transport and Social Exclusion," 207-219; Kuttler and Moraglio, *Re-Thinking Mobility Poverty.*

The workers in this study range from Dutch men and women living in urban or peri-urban areas near industries, to Belgian and German cross-border workers, migrant workers, and Moluccan exiles. These lower income and unskilled workers generally had limited resources to overcome adversity. By studying urban, peri-urban, rural, and migrant workers on their way to the workplace, *No Bicycle, No Bus, No Job* moves beyond the focus on urban (working-class) road users, common in mobility historiography.

I focus on these industrial growth centers to analyze how workers gained employment in widely varying geographical locations, ranging from nearby urban housing and peri-urban areas to remote rural villages and isolated housing sites for migrant workers. These areas differed in terms of spatial distance from industrial centers and socioeconomic geography. Many rural and migrant workers who landed jobs in Dutch industries came from impoverished regions with little prospect of work: they were forced to work further afield and travel significant distances, often with limited or poor mobility options. Yet it is important to bear in mind that these jobs usually provided a better and steady income: men earned more as a miner than as a farm or factory laborer, and women otherwise reliant on domestic labor enjoyed better pay in electronics and textile factories.

The study focuses on the actions of governments, industrial employers, and manual workers—occasionally reflecting on trade unions. According to Kuttler and Maraglio, mobility poverty research typically focuses on deprived social groups, governments that (fail to) provide bicycle paths, roads or public transit, and occasionally civil society groups that represent underserved populations or subaltern forms of mobility like walking and cycling.[56] Generally, employers fall outside this scope, though historically they were involved in many aspects of workers' daily lives as we have seen. Workers and employers—overlooked by mobility historians but identified as powerful agents by labor historians—exercised governance power in combination with the state. Including employers' and workers' perspectives in the study of commuting and job accessibility, reveals governing mobility was—and is—not solely a practice or responsibility of state actors or mobility-related advocacy groups. As *No Bicycle, No Bus, No Job* argues, workers and employers faced a similar dilemma: without efficient and affordable mobility, workers had no job and employers no workers. And since the late 1960s, scholars have argued that this is an issue the state should address.

56 Kuttler and Moraglio, *Re-Thinking Mobility Poverty*, 14.

Grasping the Worker's Perspective of Mobility

The worker's perspective provides insight into mobility systems within their historical context. As historians know too well, and mobility historians like Gijs Mom, Colin Divall, and Peter Lyth have repeated, the challenge for a history-from-below is that ordinary people's experiences left barely any traces: "it requires ingenuity, tenacity and often no small measure of luck in conceiving of and locating sources that allow one to grasp something like the full complexity of human mobility."[57] Indeed the search for useful source materials has been challenging, but not impossible. In Dutch labor party and trade union archives at the International Institute for Social History in Amsterdam, and the Dutch Diary Archive (Nederlands Dagboekarchief), the everyday commute is hardly mentioned at all. With this absence of direct sources, I searched and found echoes of workers' voices in sources provided by employers, chambers of commerce, local governments, and stakeholders likely involved in workers' (im)mobility.

It is not only challenging to find sources, but also to avoid reading them exclusively and uncritically from the perspective of those in power. Since I am interested in industrial-capitalist employers' mindset, and question the underlying power relations, politics of control, and capitalist interests, I read source material in a way that theorists call, *along* and *against* the grain: to analyze the dominant reading of a text and engage in alternative or "resistant" reading. Such a reading scrutinizes the beliefs and attitudes that typically go unexamined in a text. It draws attention to the sources' gaps, silences, and contradictions.[58] Because of my interest in the workers' perspective, I often had to read between the lines and combine various historical and secondary sources to paint a fuller picture.

Evidence presented in this thesis ranges from primary and secondary written sources to quantitative data, which provide clues and information about mobility barriers and policy decisions in the past. First, searching digital newspaper and journal databases helped locate events and historical actors. Historical newspapers are available online in the Koninklijke Bibliotheek Delpher Database and via historical centers in Eindhoven, Rotterdam, and Schiedam. This processing identified moments in time

57 Mom, Divall, and Lyth, "Towards a Paradigm," 33.

58 Reading against the grain is a methodological approach to (archival) sources developed by feminist scholars like Jamie Berlowe-Kayes in "Reading against the Grain: The Powers and Limits of Feminist Criticism of American Narratives," *The Journal of Narrative Technique* 19, no. 1 (1989): 130-140; and later by historical anthropogist Ann Laura Stoler, see *Along the Archival Grain: Epistemic Anxieties and Colonial Common Sense* (Princeton: Princeton University Press, 2009).

when workers faced barriers in accessing their places of work, providing windows to seek archival sources.

Second, for a more aggregate picture of evolving commuter patterns, I relied on small but rich historical scholarship on commuting, like the dissertations by Gerardus Theodorus Jozef Delfgaauw *De tendenzen tot decentralisatie in de vestiging der nijverheid* (1932) and Frits Bakker Schut *Industrie en Woningbouw* (1933). Census reports for 1947, 1960, and 1971 provide data on commuter patterns in terms of origin-destination, modal split, and occasionally class, gender, and age. These sources helped sketch a broader picture of mobility patterns and historically changing living and working locations.

Third, I consulted the Enschede, Eindhoven, and Rotterdam city archives, as well as Vaals archives, the Limburg Regional Historical Center in Maastricht that holds the State Mines and Oranje Nassau Mines archives, the National Archives, and NIOD Institute for War, Holocaust and Genocide Studies. These collections include city council minutes, correspondence, and reports, company, chambers of commerce, and factories' archives, and Ministry of Public Works collections. In the case of textile industries in and around Enschede, postwar bus transport was organized by an overarching organization: Central Bureau for Industrial Personnel Transportation (Centraal Bureau Industrieel Personeelsvervoer, CBIPE) and its collection is kept in Enschede's municipal archive. For the war years, NIOD collections provided valuable information about the negotiations between employers and the occupying Nazi-government, as well as the implementation and impact of centralized austerity measures on workers' everyday mobility.

A fourth source is company correspondence, internal minutes, reports, social affairs, and transportation departments. A novel aspect of my research is that I used these collections to gain insight in mobility development paths. Large Dutch companies' documents are not publicly accessible, like Philips Company Archives (PCA) and Tata Steel Central Archives (TSCA, formerly known as Hoogovens). Here I relied on the expertise and resourcefulness of professional company archivists who found relevant sources regarding the (in)direct role employers played in governing workers' mobility. Correspondence between workers, employers, state and non-state actors revealed the negotiations and networks underlying the governance of workers' mobility.

A fifth rich source is company personnel magazines. The International Institute for Social History houses a large collection. The in-house magazines I used are electronics company *Philips Koerier*, Hoogovens *Samen*, and textile industry magazines *Spil en Spoel* and *Mero-Meningen*. State Mine personnel magazines *Stukkool* (1929-1942), *Steenkool* (1946-1955), and *Nieuws van de Staatsmijnen* (1952-1975) are available via heritage website De Mijnen, and personnel magazine *Wilton-Fijenoord Nieuws* via Schiedam city council archives. These magazines provide valuable insights in employers' perspectives on workers' mobility, as well as the orchestrated collaborations with other social actors to govern bicycle and moped riding workers.

For the 1970s, when the mining and textile industries collapsed, similar sources were not available, except for Hoogovens and Philips. For this period, work by Dutch sociologist Enne de Boer, who translated international scholarship on what was then referred to as 'transport poverty' into a Dutch context, proved invaluable, along with reports and magazines from Dutch bicycle and public transit advocacy groups like the Cyclist's Union and ROVER, established mid-1970s. Combined with newspaper items on transport poverty and job access debates during the 1970s recession and a collection of jurisprudence *Passende Arbeid* (Suitable Work), these materials gave me a sense of how car-less, captive cyclists and bus riders were able to get to work.

To understand the changing meanings of workers' mobility, *No Bicycle, No Bus, No Job* treats the topic chronologically. Chapter 1, "Responding to the Transport Mismatch, 1920-1940," highlights that workers opted for mobility alternatives to rail-based options. In this interwar period marked by recession, widespread unemployment, and a growing mismatch between where people lived and worked, cheaper bicycles and buses were vital for landing jobs. At the same time, these modes also became increasingly scrutinized by government authorities. In the case of paratransit bus services, Chapter 2, "Protesting Bus Regulations during the Depression, 1926-1938," shows how workers took action when the option to commute by bus was severely curtailed. In a case study of the Limburg mining region, I reveal the social impact and miners' active resistance against these bus regulations. With these first two chapters, I show that the state provided the physical and legal infrastructures that set the boundaries for how people could move, yet workers, eager to earn a living during the Great Depression, were resourceful and resilient in shaping their everyday mobility. Chapter 3, "Mobility Austerity during War and Scarcity, 1940-1947," covering the World War II period and its direct aftermath, reveals this was a precarious way of life. War efforts and widespread shortages raised mobility barriers. The German occupier, in co-operation with industrial employers, imposed national mobility austerity measures. In effect, the locus of control over mobility shifted from workers to the state and employers—serving as a prelude to the postwar era.

Historiography describes the postwar era as a "mobility explosion," due to the greater availability of new (motorized) transport technologies. In chapters 4 and 5, I show that the practice of commuting greater distances was also born out of necessity. The Netherlands struggled with severe housing shortages until 1960 and public transit failed to bring solace for long-distance commutes as Chapter 4, "Mobility Barriers during Postwar Industrialization, 1947-1970" shows. This was also acknowledged by employers. Chapter 5, "Postwar Mobility Practices, 1947-1970," shows how blue-collar workers' travel horizons varied depending on their homes' location and available mobility options. This chapter details how urban and peri-urban workers typically cycled to their workplace—some discovering mopeds by the late 1950s—but rural and migrant workers travelled up to 100 km distances by company bus. Amid postwar industrialization, workers' mobility—like housing had been for decades—became a domain of company intervention, Chapters 6 and 7 reveal. Chapter 6, "Disciplining Cyclists and Moped Riders," details that as increasing motorized traffic made postwar roads busier than ever, company managers feared that traffic injuries and fatalities were a threat to productivity. Bicycle and moped riders were subjected to more collaboratively orchestrated forms of disciplining to prevent them falling victim to traffic. Ununionized rural and immigrant workers' movements by bus, in contrast, were top-down controlled by employers, as Chapter 7, "Mobilizing Rural and Immigrant Workers by Company Bus," explains.

Finally, the first postwar decades had seen large investments in heavy industry, a stringent government wage policy, and push for full employment: The downturn of the 1970s ended this trend. Chapter 8, "Leaving Workers to their Own Devices during Deindustrialization, 1970-1990," shows this also affected workers' mobility. With industrial closures, outsourcing, and automatization, company support for workers' mobility waned. Confronted with rising employment and forced cuts in public spending, the government gradually withdrew from supporting already insufficient public transit. Critical scholars raised concerns about the disastrous effects of decades of car-centered transportation and land-use planning on the car-less population's ability to travel—illustrated by the carpenter's case. Car ownership and usage had increased dramatically since the 1960s. For many workers, however, it was an enforced choice due to the lack of alternatives for traveling the great distances between home and workplace. For those without a car, a more limited travel horizon became a daily reality. And here we have come full circle. This concluding chapter describes the 1970s as a pivotal moment for scholars thinking about (im)mobility. It stands as a pre-amble to today's mobility poverty and justice debate.

1 Responding to the Transport Mismatch, 1920-1940

A photo (fig. 2) taken in the summer of 1929, shows workers leaving a Philips electronics factory site in Eindhoven on foot or by bicycle. A passenger bus is carefully navigating the crowded streets and there is a car in the background. Until the 1960s, only the well-to-do owned a private car, so this was likely driven by an engineer or higher-up staff. At a time when the state had been constructing roads for future automobility, most blue-collar workers relied on other mobility options for their commute.

Figure 2: Workers leaving Philips factories on foot, by bicycle, and bus. Eindhoven, 1929 (Source: Philips Company Archives)

The vast number of workers leaving the factory elicited many comments. In 1932, *Limburger Koerier*, a newspaper in the southern Dutch mining

Bek, Patrick, *No Bicycle, No Bus, No Job: The Making of Workers' Mobility in the Netherlands, 1920-1990*. Taylor & Francis Group 2022
DOI: 10.5117/9789463723183_CH01

region, marveled at the scale and dynamics of workers' daily travel to the town Geleen and state mine Maurits, the largest in the area. The editor described the mine as the central point of the wider region: "Just like the radial threads of a spider web leading to the center ... From all sides the lines along which the daily traffic moves to this agglomeration come together there." Not only the import and export of coal, cokes, mine wood and other goods, but also "the human material earning its living in the dark bowels of the earth or on the extensive sites above ground with their workshops, their coking plants, their offices, etc." The editor highlighted the continuous rhythm of thousands of bodies moving in one direction towards the mines over unchanging routes, by comparing miners to "industrious bees or active ants that continuously seek their hive, their nest." Most local (urban) and unskilled workers walked to the pit, while the better paid miners from further away came by bicycle, bus, or rail.[1]

The issue of how workers got to work was not new in the 1920s, but was more broadly picked up by governments and engineers across Europe, as Colin Divall and other historians have noted in passing.[2] The relocation of industries and the need for more cheap labor for expanding industries, confronted workers, employers, and governments with how to best overcome the spatial mismatch between home and work locations.[3] Many labor historians have shown that industrial-capitalist employers were driving forces in housing construction, seen as an attempt to reduce spatial distances between cheap labor forces and factories or mines.[4] But since only small segments of workers were actually accommodated in company housing, the majority of employees had to travel to work every day. "Responding to the Transport Mismatch" shows that government and employers' solutions—railway and housing construction—fell short. Workers looked elsewhere: they found affordable solutions by opting for bicycles and buses.

1 "De mijnwerker en z'n fiets," *Extra Uitgave van de Limburger Koerier*, 30 June 1932, 17.

2 Colin Divall and Winstan Bond, *Suburbanizing the Masses: Public Transport and Urban Development in Historical Perspective* (London/New York: Routledge, 2003).

3 The "spatial mismatch hypothesis" was first coined by American economist John Kain. In 1968, he described how low-income African-American jobseekers could not land jobs because affordable housing was located far from jobs and mobility options to overcome this mismatch were lacking. This mismatch led to high unemployment among inner-city jobseekers. John Kain, "Housing Segregation, Negro Employment, and Metropolitan Decentralization," *Quarterly Journal of Economics* 82, no. 2 (1968): 175-197.

4 Oliver Dinius and Angela Vergara, *Company Towns in the Americas: Landscape, Power, and Working-Class Communities* (Athens, GA: University of Georgia Press, 2011); Sjef Stoop, *De sociale fabriek: Sociale politiek bij Philips Eindhoven, Bayer Leverkusen en Hoogovens IJmuiden* (Utrecht: Stenfert Kroese, 1992).

"Responding to the Transport Mismatch" reveals two major responses to the growing transport mismatch in the interwar period, not by the state or employers, but by self-governing workers. During the 1920s and even more so during the Great Depression, workers individually appropriated bicycles as utility rather than as leisure vehicles, enabling them to move around freely, land jobs, and make a living. Bicycles had been used by the well-to-do for sightseeing. Using bicycles for commuting was novel. It represented a major innovation through a process of repurposing.[5] There were also more organized and collective responses to the transport mismatch. Both workers and employers discovered motorized passenger buses as a more flexible form of transport. Like elsewhere in Europe, these paratransit buses operated within an existing transport system that failed to meet workers' needs. Transit users rather than authorities innovated the services in a bottom-up process. Not only local entrepreneurs, but blue-collar workers and some employers organized grass-roots passenger buses, as we see in the next section.

1.1 Transport Mismatch between Home and Work

The transport mismatch challenges occurred in the wider social and economic context, regarding workers' ability to get to work. The interwar period was a time of industrial expansion. In his analysis of GDP in the Netherlands between 1889 and 1990, economic historian Harmen Jan de Jong shows nadirs of industrial production during both World Wars and general production growth (about 3.78 percent; GDP 2.95 percent per year) in the interwar period, despite the 1930s economic recession.[6] New industries emerged (steelworks Hoogovens), and existing industries—like lightbulb manufacturer Philips, the Twente region's spinning, weaving and bleaching factories, and the province Limburg's private and state mining companies—expanded operations and increased productivity. The reasons for this economic expansion included government support for growth industries in a push for economic autonomy after the First World War, the introduction of scientific management, and the mechanization of manufacturing processes.[7] Mobilizing enough workers to match the expansion, however, remained key for industrial productivity until the 1960s.

5 For a discussion on how users repurposed technologies, see Ruth Oldenziel and Mikael Hård, *Consumers, Users, Rebels: The People who Shaped Europe* (London: Palgrave, 2013).

6 Harmen Jan de Jong, *De Nederlandse industrie 1913-1965: Een vergelijkende analyse op basis van productiestatistieken* (PhD Dissertation, Rijksuniversiteit Groningen, 1999), 2-4.

7 Frits Bakker Schut, *Industrie en woningbouw: Een technisch-economische beschouwing over bemoeiingen van de industrie met arbeiderswoningbouw* (Assen: Van Gorcum & Comp., 1933), 20.

Industries' reliance on low wages partly explains the increasing mismatch between workers' home and workplace. As industries expanded and local labor markets could no longer meet the increasing demand, employers had to attract cheap labor from further away. For rural workers in domestic production and agriculture, this meant new job opportunities in Dutch growth industries. Workers from the countryside could only realize such opportunities if they were able to reach these places. Initially, the expansion of national and regional railways, and local tramways improved the mobility options between workers' homes and (suburban) industrial sites.

Labor historians have observed a commuter trend across Europe. Since the 1890s, Flemish people working in the French textile industry did not move to France, but preferred to stay in Belgium, where the cost of living was lower and they could commute daily. In the late nineteenth century, South Wales and the Great Northern Coalfields railways in Britain unlocked the pits for workers commuting from the wider region. By the early 1900s, Polish workers migrated but also commuted daily or weekly to manufacturing jobs in Germany. Across interwar Britain, former agricultural workers commuted from their villages with dead-end jobs and few prospects to regional industrial centers. In Belgium's South-West Flanders region, rail expansion was instrumental in unlocking distant industrial jobs for rural workers. And according to historian Serge Langeweg, railway expansion was instrumental to industrial growth and labor market expansion in the Dutch mining region.[8]

Two other developments aggravated the transport mismatch: as cities and towns expanded outward, industries relocated to the outskirts and isolated places. The industrial relocation to locations outside urban centers led to both the decentralization and expansion of labor markets. Dutch economist

8 Robert Baker, "Socialism in the Nord, 1880–1914. A Regional View of the French Socialist Movement," *The International Review of Social History* 12, no. 3 (1967): 357-389; Greet De Block, "Urbanizing the Countryside: Rails, Workers and Commuting in South-West Flanders, Belgium, 1830-1930," in *Cultural Histories of Sociabilities, Spaces and Mobilities*, ed. Colin Divall (London: Routledge, 2015), 53-66; Simon Constantine, "Migrant Labour in the German Countryside: Agency and Protest, 1890–1923," *Labor History* 47, no. 3 (2006): 319-341; Martin Daunton, "Miners' Houses: South Wales and the Great Nortern Coalfield, 1880-1914," *The International Review of Social History* 25, no. 2 (1980): 143-175; Ian Kerr, "On the Move: Circulating Labor in Pre-Colonial, Colonial, and Post-Colonial India," *The International Review of Social History* 51, Supplement 14 (2006): 85-109; Serge Langeweg, *Mijnbouw en arbeidsmarkt in Nederlands-Limburg: Herkomst, werving, mobiliteit en binding van mijnwerkers tussen 1900 en 1965* (Hilversum: Verloren, 2011), 63-72; Dhiraj Nite, "Employee Benefits, Migration and Social Struggles: An Indian Coalfield, 1895-1970," *Labor History* 60, no. 4 (2019): 372-391; Randall Patton, "Textile Organizing in a Sunbelt South Community: Northwest Georgia's Carpet Industry in the Early 1960s," *Labor History* 39, no. 3 (1998): 291-309; Peter Scott, "Women, Other 'Fresh Workers', and the New Manufacturing Workforce of Interwar Britain," *The International Review of Social History* 45, no. 3 (2000): 449-474.

G.Th.J. Delfgaauw already noted the trend in 1932. Although nineteenth-century industrialized cities attracted new industries, the economist observed "the strong trends towards decentralization" in the Netherlands and elsewhere in Europe by the early 1900s. The "social issues" and changing public discourse regarding urban crowding motivated decentralization in land-use planning. Garden City associations in Britain and Germany sought to develop strategies promoting the relocation of industries to the urban fringes, combined with separate healthy, livable towns and villages.[9] This international movement also led to the construction of company housing and garden villages across industrializing Netherlands—like Philipsdorp in Eindhoven for former peat cutters from Drenthe employed at Philips, and Heijplaat on the south banks of Maas River in Rotterdam, housing shipyard RDM dockworkers.[10]

According to Delfgaauw, new communication and (affordable) transportation technologies enabled large-scale industries to escape the higher city center prices and municipal leaseholds. Small industries often remained in cities as did companies like Philips, which stayed in Eindhoven. The Rotterdam Port Authority (Rotterdams Havenbedrijf) expanded its harbor towards Schiedam and Vlaardingen because no land was available in Rotterdam; shipyard RDM went for the isolated south riverbank and Wilton-Fijenoord moved part of their docks to Schiedam. Steel company Hoogovens, despite an option to locate its new blast furnaces in Rotterdam, finally settled at the more isolated Noordzeekanaal estuary at IJmuiden, where there were no significant industries or labor reserves, but land was cheaper and soil firmer.[11] Proximity to (labor) markets, transport hubs, financial institutions, and other companies was essential. In heavy industries like mining, closeness to natural resources (coal fields) determined the location. For other manufacturing industries, proximity to (labor) markets no longer guided their choice of location. With the emergence of communication technologies

9 Alfred Weber's model of industrial location came down to: first a location with the lowest transportation costs, then labor costs and agglomeration economies, i.e. mutual attraction and clustering of different industries in one place. In the early 1900s, industries clustered near consumer markets and transport nodes and new sites emerged around urbanized centers. Alfred Weber, *The Theory of Location of Industries* (Chicago: University of Chicago Press, 1929 [1909]); Bas van Leeuwen, Robin Philips, and Erik Buyst, *An Economic History of Regional Industrialization* (London: Routledge, 2021).

10 Len de Klerk and Ries van der Wouden, *Ruimtelijke ordening: Geschiedenis van de stedelijke en regionale planning in Nederland, 1200-nu* (Rotterdam: nai010, 2021), 108-113.

11 Gerardus Delfgaauw, *De tendenzen tot decentralisatie in de vestiging der nijverheid*, Serie voor Stedebouw VII (Amsterdam: Nederlandsch Instituut voor Volkshuisvesting en Stedebouw, 1932), 17-20.

like intercommunal telephony and faster mobility options, proximity was no longer a decisive factor.[12]

Delfgaauw collected data in 1926 showing how 11 percent of the workers in Eindhoven commuted from the surrounding region, even from Belgium, and that 20 percent of Twente's textile workers lived in other towns and villages, also across the German border. The number of commuters in Limburg province was unprecedented in the Netherlands: over a third of the working population travelled from remote villages and across the Belgian border to the mines. Around 6,500 miners travelled daily to the mining town Heerlen, the region's industrial center, accounting for almost half of Heerlen's working population.[13] The ability to travel these distances or settle near workplaces had always been essential for getting or keeping a job. Decentralization posed a challenge for both workers and employers. The government and employers centered on two predominant solutions: expanding the railways and company housing.

Dutch civil engineer Frits Bakker Schut, later head of Philips' housing department, argued in his 1933 dissertation *Industrie en Woningbouw* (Industry and Housing) that company housing should be an integral part of the industrial production process. Since then, historians have shown that paternalistic employers constructed housing in an effort to stabilize their labor force and avoid the unbridled labor turnover and absenteeism, by imposing a higher degree of control over workers and reducing the physical travel distance between workers' homes and the work floor. Because these company housing schemes were near factories or mines, workers could easily walk or cycle to work. Company housing thus offered some solace. Yet, employers found building company housing was a costly undertaking. Moreover, regional workers preferred their village homes over company housing where migrant workers—from elsewhere or across the border—typically stayed. In the nineteenth century already, international studies showed that too much interference by employers in workers' lives had a counter effect. Company housing in particular made many workers feel they were always within their employers' reach, even in the confines of their own home.[14] Company housing was thus not a fix for all.

12 Delfgaauw observed that transporting workers over greater distances involved expanding labor markets. Workers could be attracted from the countryside and industries in remote locations could attract urban workers. *De tendenzen tot decentralisatie in de vestiging der nijverheid*, 13, 15, 17-18, 21.

13 *De tendenzen tot decentralisatie in de vestiging der nijverheid*, 53-55, Appendix I-VI.

14 Bakker Schut, *Industrie en woningbouw*, 23, 27-30, 34, 41-42, 81; Ad Otten and Elisabeth Klijn, *Philips' woningbouw 1900-1990: fundament van Woningstichting Hertog Hendrik van Lotharingen* (Zaltbommel: Europese Bibliotheek, 1991), 215; Stoop, *De sociale fabriek*, 55-56.

The state and companies also believed that expanding the railways would solve this spatial mismatch. With industrial growth and the relocation of industries and residential areas to urban outskirts, mobility historians show that rail transport infrastructures had to accommodate workers' mobility over longer non-walkable distances. In Italy, Turin's political and economic elite favored a local train network to improve their economic position and mobilize the thousands of workers in a typical factory town.[15] Rail commuting became common practice across Europe. In fin-de-siecle London, around 325,000 workers travelled via suburban railways, paying reduced fares.[16] In Berlin, the Ringbahn was the main mode of transport for workers living in southern and eastern districts like Rixdorf and Friedrichshain, who commuted to the north-western industrial areas Wedding and Spandau. Trains enabled workers to travel from their suburbs in the early morning to work, in time for the morning shift. Travel time was reduced. Instead of having to walk an hour or more, workers from Hechtsheim, 7 km south of Mainz, could now get back to their village within half an hour. A train trip was also affordable: a one-way ticket cost 30 pfennig for a 10 km ride. This was about 8 percent of their daily wage, but still outweighed cheaper rents and agrarian subsistence in rural villages.[17] Similar developments in Belgium's South-West Flanders region enabled rural workers to access better paid jobs further afield.[18]

Rail expansion proved instrumental, not just in creating a mobile labor force. According to Greet De Block, the new rail networks, cheap workmen fares, and adapting timetables to workers' shifts, helped to expand the Southwest Flanders' labor market. These measures also effectuated "the emancipation of the working class," in an economically liberal sense, providing material equality to laborers isolated in rural areas, notably for going out to pursue their interests and needs.[19] Tomas Ekman earlier argued about trains: "cheap enough for the ordinary worker to use and should make

15 Massimo Moraglio, "Rails in the Car Kingdom: Competition, Changes and Continuity in Urban Mobility – The Case of Turin, 1914-1973," in *From Rail to Road and Back Again? A Century of Transport Competition and Interdependency*, ed. Ralf Roth and Colin Divall (London/New York: Routledge, 2015): 355-372, here 369-370.

16 Paolo Capuzzo, "Transport and (sub)urban development: Between politics and technology: transport as a factor of mass suburbanization in Europe, 1890-1939," in *Suburbanizing the Masses*, ed. Divall and Bond, 25-26.

17 Dieter Schott, "Suburbanizing the Masses for Profit or Welfare: Conflict and Cooperation Between Private and Municipal Interests in German Cities, 1890-1914," in *Suburbanizing the Masses*, ed. Divall and Bond, 85-86.

18 De Block, "Urbanizing the Countryside," 53-66.

19 "Urbanizing the Countryside," 53-55, 58.

it possible for them to leave the dark and dirty city behind and move out to the new suburbs in the countryside."[20] Bond shows that these notions, fueled by social ideas about welfare and deproletarization—and fears of labor unrest and turnover, were a widespread ideology in western, industrialized countries: "The goal of better housing for the working classes away from the smoke, disease, and congestion was a shared dream of reformers on both sides of the Atlantic in the early 20th century."[21] In Germany too, according to Schott, the late nineteenth century expansion of suburban railways in industrial places like Mainz gave workers a better and healthier life, while rural workers could "spend one hour more with their families every day."[22] Reduced mobility barriers to trams and trains enabled working-class people to commute and live in the suburbs or the countryside where rents were cheaper.

New mobility measures and innovations enabled commuting. In *Suburbanizing the Masses* (2003), historian Paolo Capuzzo shows how tramway companies and city governments in Europe encouraged workers to use rail-based modes for home-work trips. Expanding rail networks from working-class neighborhoods to (sub)urban industrial workplaces was one solution. Reducing workmen fares was a "very important measure of industrial policy," enabling labor market expansion for growth industries.[23] In the Netherlands too, reducing cost barriers to rail-based mobility was common practice. In *Industrie en Woningbouw*, Bakker Schut mentioned that cheap fares had existed for Dutch workers since the late 1800s. Lower working-class commuters travelled in third or fourth-class carriages, segregated from upper-class passengers.[24]

The Dutch government supported rail connections through interest-free loans for railway construction. Langeweg shows how shortly after the First World War, the Dutch Ministry of Public Works supported tramway and railway expansion to unlock labor reserves for the mining industry. Previously, state mine supervisors (Staatstoezicht op de Mijnen) complained about the Limburg region's poor mobility options. It took miners hours to reach the mines not only because limited railway tracks forced them to take time-consuming detours, but the few stations meant they had to walk several kilometers to the pits. Also, train timetables were not adjusted

20 Tomas Ekman, "Vision in Solid Form: A Comparison Between Two Solutions to the Traffic Probelm in Stockholm, 1941 and 1992," in *Suburbanizing the Masses*, ed. Divall and Bond, 181.

21 Winstan Bond, "The Flawed Economics and Morality of the American Uniform Five-cent Fare," in *Suburbanizing the Masses* ed. Divall and Bond, 65.

22 Schott, "Suburbanizing the Masses for Profit or Welfare," 79-80.

23 Capuzzo, "Transport and (sub)urban development," 25-26.

24 Bakker Schut, *Industrie en woningbouw*, 81.

to mining shifts, leaving miners on the night shift with no homebound transport.[25] According to a 1914 report in engineering journal *De Ingenieur*, the Dutch parliament found transit rail for workers crucial for industrial growth regions like Southeast Limburg. Trains and trams were considered a "comfortable mode of transport" for workers "who thus far had not been inclined to land a job at the mines because of the distance to their home, but now would be eager to do so."[26] In 1916, engineer A.M. Harthoorn argued in *De Ingenieur* that railway expansion had indeed created a mobile labor force: "The miners come ... from all towns and villages between Maastricht and Roermond in the south of Limburg, ... workers' transportation there is extensive along all routes."[27]

Trains helped mobilize workers in other places. Local Eindhoven newspapers describe how an additional platform was built in 1929 to cater for the rapidly growing number of workers getting on and off the trains at Hagekamp near Julianastraat. Around the same time, Strijp station was opened for electronics factory Philips staff commuting from Belgium. At another new station Philipsdorp Zuid, the "Philips Express" ran services in the morning and evening, bringing "large contingents of Belgian workers." The electrification of the rail tracks between Haarlem and IJmuiden by Dutch Railways (NS) unlocked the relatively isolated steel company Hoogovens. IJmuiden-Oost station opened in 1925 and two additional stations near the blast furnaces Velsen-Hoogovens and Koog Bloemwijk. In Twente, regional newspaper *Tubantia* reported that textile factory managers successfully lobbied for a train stop in the Oldenzaalschestraat between Enschede and the village Glanerbrug, improving rail access for workers living in Enschede's rural surroundings.[28] These rail connections helped employers to secure a cheaper labor force.

Dutch industrial employers guided these rail mobility developments. They lobbied with state and Dutch Railways for workmen fares, adjustment

25 Langeweg, *Mijnbouw en arbeidsmarkt*, 64-66.

26 "Aanleg van een spoorweg Sittard-Heerlen-Pruisische grens," *De Ingenieur* 29, no. 26 (1914): 511-512.

27 A.M. Harthoorn, "Is het wenschelijk den locaalspoorweg Sittard-Herzogenrath tot hoofdspoorweg te maken," *De Ingenieur* 31, no. 22 (1916): 413-415.

28 "Twentsch Nieuws," *Twentsch Dagblad Tubantia*, 28 April 1916, 2; "Binnenland," *Het Vaderland*, 4 October 1925, 1; "Extra treinen voor de Philips-arbeiders," *Eindhovensch Dagblad*, 10 July 1929, 1; "De werkliedentreinen naar Eindhoven," *Eindhovensch Dagblad*, 7 June 1930, 1; "Textiel in Twente," Mobiliteits Museum, Accessed June 18, 2021, http://www.mobiel-erfgoed.nl/VMM_TextielTwente_Intro.htm; H. de Jager, "Personenvervoer bij de Hoogovens," *Rail Magazine*, no. 174 (2000): 20-21; S. Overbosch, "Stopzetting arbeidersvervoer Enschede-Glanerbrug," *Op de Rails*, no. 4 (1951): 25-26.

of timetables, and extra worker trains—all aimed at mobilizing labor. Employers even paid employees modest travel allowances. The Philips human resources department partially reimbursed workers' travel costs based on weekly wages. These reimbursements were not necessarily based on need but on status: those on higher salaries received more. In 1929, a Philips employee earning between 6 and 13.99 guilders a week received a weekly travel allowance of 1.60, whereas an employee earning 27 guilders received 5.05 guilders. These allowances were discretionary due to the high costs for employers, especially in the Netherlands, where rail travel was more expensive than in neighboring countries.[29]

Neither company housing nor railways, however, could fundamentally resolve the spatial mismatch. Notwithstanding employers' involvement in rail-mobility, workers found that the promise of rail-based mobilization worked out differently. If workers did not live near enough to walk to work, they depended on expensive, inflexible, and often uncomfortable worker trains. Rail exploitation, moreover, was a costly undertaking for the state and rail companies. Land purchase, then the construction and maintenance of rails, tunnels, and bridges, demanded large investments, and return on investments became increasingly uncertain in the 1920s. For workers relying on trams and trains, the travel costs rose significantly over time. State-supported rail transit neither met workers' or employers' mobility needs.

Workmen rail passes were common practice but not consistent. In November 1921, State Mine and Oranje Nassau Mines management wrote to Dutch Railways and the Ministers of Agriculture, Industry and Trade (Landbouw, Nijverheid en Handel) and Public Works Administration to express their dismay about the announced discontinuation of workmen passes in January 1922. The mine operators felt that Dutch Railways should not shift the burden of declining revenue onto low-income workers, and considered the imminent fare increase especially problematic at a time of plummeting coal prices and thus lower wages. They asked Dutch Railways to uphold the special fares for miners.[30]

Philips also considered the cost of rail-based mobility generally too high for its manual workers. In 1929, Philips wrote to the Minister of Labor, Trade, and Industry, accusing the Railways of ignoring the development of

29 Philips Company Archives (PCA hereafter), Inventory no. 631.4, no. 1, Correspondence April 12, 1929.

30 Regionaal Historisch Centrum Limburg (RHCL hereafter), Archive no. 17.05H, Oranje-Nassau Mijnen, 1903-1978, Inv. no. 18, Arbeidersvervoer per trein, 1905-1949, Correspondence 19 November 1921.

trade and industry and exploiting manual workers. Dutch workers living forty kilometers from factories paid 20 guilders a month for a rail pass, compared to 5.30 in neighboring Belgium and 13 in Germany.[31] These fares were a serious expense for a household of poorly paid manual workers. Most manual workers were unskilled or semi-skilled, often without a permanent employment contract, who got paid daily, weekly, or a piece wage. Wage levels varied per branch, skill level, region, gender, and age. In the mid-1930s, men in the manufacturing industries earned between 54 and 63 cents an hour in Rotterdam, and between 36 and 52 cents in Eindhoven. Overall, women earned significantly less, between 10 and 24 cents, increasing with age. Economic developments determined wage levels. During the 1930s economic downturn, prompting mass lay-offs and wage cuts, the average weekly pay for a man working in Dutch manufacturing dropped to 21.50 guilders in 1936 (compared to 29 in 1920). For those who lost their jobs and had to rely on government welfare schemes, income dropped by 40 to 50 percent.[32] During the Great Depression, the high cost of rail travel weighed heavily on low-income workers and jobseekers.

Dutch employers organized their opposition to the high workmen fares. The Association of Dutch Employers (Verbond van Nederlandsche Werkgevers), the organization representing industrial companies in the Netherlands, considered third-class workmen fares of 60 cents per km too high. In 1929, while the Railways had reduced fares for first and second-class passengers from 1.20 and 0.90 to 1.08 and 0.86 guilders, they did not lower the workmen

31 PCA, no. 642.5, Personeelsvervoer, Inv. no. 2, Correspondence 29 March 1929, 2-4, and 4 April 1929, 1-2. By reducing worker train tariffs, workers no longer had to live packed in urban industrial areas. Workers would not be "exposed to temptations of city life, and their proximity to kin and low-cost rural life were arguments for commuting. Such deproletarization arguments were important reasons for keeping train travel affordable for workers. Philips claimed that Dutch Railways ignored these social and economic aspects, and approached workmen fares "from a sole cost-benefit perspective" ("een zuiver exploitatie oogpunt").

32 Centraal Bureau voor de Statistiek, *Statisch Zakboek 1935* (The Hague: Martinus Nijhoff, 1935), 29; Centraal Bureau voor de Statistiek, *Statisch Zakboek 1938* (The Hague: Martinus Nijhoff, 1938), 25; Peter Schrage, Erik Nijhof, and Piet Wielsma, "Inkomensontwikkeling van werkenden en werklozen in Nederland (1913-1939)," *Tijdschrift voor Sociale Geschiedenis* 14 (1989): 347-394, here 374-375; Jeroen van der Spek, "Een eeuw lonen en prijzen (1870-1970)," *Sociale Maandstatistiek* 19 (1971): 418-426, here 421. Household incomes are more difficult to determine: if only the male breadwinner or other household members performed waged work. Wages differed per branch and fluctuated with conjectural developments. CBS data indicates that in 1921, a metal worker earned a *daily* wage of 0.77 guilders, in the 1930s between 0.53 (1937) and 0.70 (1930). Miners earned 7.45 to 5.23 guilders in 1921, fluctuating between 5.05 (1935) and 5.85 (1930) for underground miners, and 3.87 (1935) and 4.28 for above ground miners. A dock worker earned a *weekly* pay of 39.39 guilders in 1921, and between 29.55 (1935) and 35.42 (1930).

fare that, in some cases, equaled an hourly wage. The Association urged the national government and Railways to provide fares low enough to enable workers to go on living in "the somewhat remote areas where they come from," where the cost of living was lower compared to industrial centers. They also argued that low fares allowed lower wages without putting rural workers at a disadvantage compared to their urban co-workers, who paid higher rents and sustenance. The high cost of mobility could not be borne by low-income manual workers, nor by industrial employers, the Association argued.[33]

Rail-based options also fell short in terms of services. In 1922, Philips complained to the Chamber of Commerce that due to limited train and tram connections to Eindhoven, it was impossible for staff to get to work on time or they had to leave work earlier to get home that day. Hundreds of workers from the villages Sint Oederode, Valkenswaard, Veldhoven, and Weert experienced these problems, causing trouble at work.[34] Similarly at Rotterdam, in December 1928, twenty dock workers at the newly developed Waalhaven docks complained in a letter to the Chambers of Commerce and Factories about the poor accessibility of the area. T.K. Diepenhorst, who lived an hour and a half's walking distance away in Rotterdam's Charlois neighborhood, found that the sole mobility option to the docks, the tram, fell short; as the frequency was too low, many dock workers had to find other means of transport. If this situation was not resolved, undersigned workers claimed: "it would cause a lot of trouble due to them arriving late for work."[35]

The last and first leg of the journey to the final destination was similarly problematic. Miners in Limburg had to walk an hour along muddy roads and over hilly terrain to the nearest train station, then after an hour or more train ride, walk from the station to the mine shaft. Researcher Bakker Schut observed in 1933 that Belgian workers at Philips in Eindhoven had to face a four-hour journey home after a twelve-hour working day and another "long march through the fields," because the train station was often far from their homes. And, he pointed out, the long travel times and shorter night's rest for workers were reasons why the company had concerns about workers' productivity.[36]

33 PCA, Archive no. 642.5, Inv. no. 1-1929.02, Correspondence 15 February 1929.

34 PCA, Archive no. 642.5, Inv. no. 1-1929.02, Correspondence 20 July 1922, 7 November 1922, 11 November 1922, 12 and 22 December 1922.

35 Nationaal Archief, Archive no. 3.17.17.04, Archief Kamer van Koophandel en Fabrieken voor Rotterdam, 1922-2001, Inv. no. 1723, Personenvervoer, Correspondence 29 December 1928, 26 January 1929.

36 Johan Blok, Cor Campagne, and Sjef Janssen, *Trams in Midden- en Zuid-Limburg 1888-1950* (Heerlen: Schuyt & Co. Uitgevers, 1998), 15; Serge Langeweg, "De geographische spreiding van

Finally, workmen trains were anything but comfortable. In the cold winter months, carriages were often unheated, especially in times of coal shortages during and after First World War, financial newspaper *Algemeen Handelsblad* noted in 1919. Mineworkers' newspapers show that miners usually had to stand in designated workmen carriages, often substituted with cattle carriages. Catholic Miner Union magazine *De Mijnwerker* recalled in 1938, the poor, almost inhuman conditions in 1920s miners' trains, nothing more than "crawling, bumping, and jolting [carriages], lacking light and air." A one-way commute took two hours or more, and after a shift, there were often no trains available to take miners home. "Three times a day hundreds of worn-out miners, half asleep, could be seen waiting for their train at Heerlen train station."[37] In short, rail-based technologies may have expanded, but workers experienced many barriers and discomfort.

These grievances created momentum for mobility alternatives. Historians have shown how railway expansion was instrumental in modern nation-building, labor market expansion, and reducing workers' mobility barriers to (distant) jobs.[38] Workers' movements had been gradually industrialized by rail-based transport since the nineteenth century.[39] The evidence here indicates that in the interwar period, however, rail travel acquired a negative association as expensive, uncomfortable, inflexible, and unreliable. Coupled with increasing commuting distances, a transport mismatch arose, serving as a driver for change. To get a job further away, workers turned to cheaper and more flexible mobility alternatives: bicycles and buses.

1.2 Workers Seek Mobility Alternatives

Labor market expansion and the decentralization of industrial workplaces together with insufficient mobility options had led to a transport

de mijnwerkersbevolking," in *Mijnwerkers in Limburg: Een Sociale Geschiedenis*, ed. Ad Knotter, et al. (Nijmegen: Van Tilt, 2012), 112-113; Bakker Schut, *Industrie en woningbouw*, 93-94.

37 "Onverwarmde mijnwerkerstreinen," *Algemeen Handelsblad*, 5 March 1919, 6; "Het vervoervraagstuk der mijnwerkers," *De Mijnwerker: Orgaan van de Ned. R.K. Mijnwerkersbond*, 12 November 1938, 362; Harthoorn, "Is het wenschelijk den locaalspoorweg Sittard-Herzogenrath tot hoofd-spoorweg te maken," 413-415.

38 See Divall and Bond eds., *Suburbanizing the Masses*.

39 Gijs Mom, Colin Divall, and Peter Lyth, "Towards a Paradigm Shift? A Decade of Transport and Mobility History," in *Mobility in History: The State of the Art in the History of Transport, Traffic and Mobility*, ed. Gijs Mom, Gordon Pirie, and Laurent Tissot (Neuchâtel: Editions Alphil, 2009): 13-40, here 29-30, 32-33; Wolfgang Schivelbusch, *The Railway Journey: The Industrialization of Time and Space in the Nineteenth Century* (Oakland, California: University of California Press, 2014 [1977]), 14.

mismatch—particularly problematic in light of industrial societies built around the assumption of a highly mobile labor force. Fending for themselves, workers and employers resourcefully and resiliently reshaped everyday mobility, as I will show. Amid economic downturn, they found affordable alternatives in bicycles and buses: these required few(er) resources, (state) investments, and large infrastructures compared to rail-based options. Low-income workers appropriated and repurposed bicycles for commuting en masse. Passenger buses developed as another bottom-up alternative.

In the early 1920s, rail expansion had promised a mobile labor force and to unlock rural working populations for both state and industry. When rail mobility fell short, the bicycle, a low-cost and self-powered vehicle, fulfilled this promise by the 1930s. Cycling historians show that workers reinvented the bicycle from a leisure device to a utility tool. In a time when walking was the norm among the working-class and public transit among the middle-class, cycling emerged as a surprising alternative. Like elsewhere around the globe, utilitarian cycling boomed in the Netherlands. In 1925, 1 out of 3.4 residents owned a bicycle. By 1936, this was 1 in 2.4 residents. In comparison: 1 out of 4.6 British residents, 3.88 in Germany, and 4 in Belgium owned a bicycle. By the 1930s, the bicycle had become a common mode of transport.[40]

Bicycles enabled workers to stay in their neighborhood or village, among kin, and gain employment in remote jobs without having to migrate to cities and company housing. In 1932, the *Limburger Koerier* observed that hundreds of miners rode their "metal steed" ("stalen ros") to work. At State Mine Maurits, you could see "them coming in long rows or in groups. From north and south, from east and west, on all roads." Before the night shift, between 9 and 10, "long trails of bicycle lights come swirling from opposite directions" from the wider region to the mines.[41]

Bicycles did not become a popular working-class mode of transport overnight. Cycling rates had been increasing since the late ninetieth century, initially for the well-to-do upper classes, but gradually within reach for low-income workers. The falling price of bicycle production was an important reason for this trend, as historians Sue Yen Tjong Tjin Tai, Frank Veraart, and Mila Davids reveal.[42] Advertisements by local bicycle retailers show how rapidly prices dropped. The cheapest Fongers bicycle model produced

40 Jan Leerink, *De verkeers-veiligheid op den weg: Een juridische, sociologische en verkeerstechnische studie (proefschrift)* (Alphen aan den Rijn: N. Samsom, 1938), 38.

41 "De mijnwerker en z'n fiets."

42 Sue-Yen Tjong Tjin Tai, Frank Veraart, and Mila Davids, "How the Netherlands became a bicycle nation: Users, firms and intermediaries, 1860–1940," *Business History* 57, no. 2 (2015): 257-289, here 19-20.

in the Netherlands, dropped from 150 guilders in 1918 to 100 in the 1920s, and 45 by the mid-1930s. Other makes like Wiecoda Rijwiel offered bicycles for even less (25 guilders), advertised in weekly socialist magazine The Worker (*De Arbeider* 1932).[43] At first, only male workers could afford a bicycle. It was also common practice among low-income workers to purchase even cheaper second-hand (often stolen) bicycles on (black) markets for prices between 7.50 and 15 guilders, according to populist newspaper *De Telegraaf* in 1939: the bicycle, "the vehicle of the poor," was "popular with those who could just about afford a bicycle."[44]

A one-time investment in a bicycle enabled a worker to travel three or four time faster than a pedestrian and more freely and cheaper than by public transport. It offered individual freedom, using no other resource than human power. *Cycling Cities* researchers therefore characterize the interwar bicycle as a "lifesaver," as it extended working-class people's action radius at a low purchase and maintenance cost.[45] Affordability explains why cycling was popular during the interwar period—not surprising given that the average manual male worker's weekly pay dropped significantly during the Great Depression, and despite unemployment subsidies, almost halved household incomes.[46]

In industrial cities and regions, cyclists surged into the streets. In Eindhoven, historic photos (fig. 3) of a rail crossing between working-class neighborhood Woensel and the Philips factories depict masses of cyclists. "As the city boomed, so did the number of cyclists in the city streets," researchers of *Cycling Cities: The European Experience* observe, estimating an 80 percent bicycle share in the 1930s.[47] Adri Albert de la Bruhèze found that in 1937,

43 "Prijzen van fietsen (Kampioen diverse jaren)," in *Fietsersbond Afdelingsarchieven: Archief en Onderzoek*, TIS Group (Eindhoven University of Technology, n.d.); "Een beroeps-fietsendief: Reeds een honderdtal karretjes gestolen," *Nieuwe Venlosche Courant*, 21 June 1930, 5; "Wiecoda Rijwiel," *De Arbeider: Sociaal-Anarchistisch Weekblad*, 24 September 1932, 4; "Diversen," *De Boerderij*, 23 June 1937, 33-36; "Diversen," *De Boerderij*, 20 July 1938, 33-34; Jos Rietveld, "Fietsgebruik in Nederland: 1880-1996," De Oude Fiets, accessed October 28, 2021, https://www.oudefiets.nl/2-algemeen/92-fietsgebruik-in-nederland-1880-1996; Rietveld, "Fongers: Geschiedenis," Fongers, accessed June 16, 2021, http://www.rijwiel.net/fongersn.htm.

44 "Fietsendieven hebben vrij spel," *De Telegraaf*, 12 February 1939, 3. Amid the depression, with 30,000 bicycles stolen per year in the Netherlands, editors considered bicycle theft "the cancer of society," *De Telegraaf* emphasized bicycles' vital importance for making ends meet.

45 Ruth Oldenziel et al., *Cycling Cities: The European Experience* (Eindhoven: Foundation for the History of Technology, 2016), 10, 12.

46 Schrage, Nijhof, and Wielsma, "Inkomensontwikkeling van werkenden en werklozen," 363-372.

47 Frank Veraart, Martin Emanuel, and Ruth Oldenziel, "Eindhoven: Engineering a Path for Bikes?" in *Cycling Cities: The European Experience*, ed. Ruth Oldenziel, et al. (Eindhoven: Foundation for the History of Technology, 2016): 53-63, here 53-54, 56.

textile city Enschede counted 20,000 cyclists at two major intersections, amounting to a 90 percent modal share (excluding pedestrians).[48] Frank Veraart and Manuel Stoffers show how in mining towns Heerlen and Kerkrade, the bicycle's share hovered around 80 percent between 1935 and 1940, placing them among Europe's leading bicycle-dominated towns.[49] In a policy report, Eric Berkers complements this research, observing bicycle shares of up to 95 percent in the 1930s in Southeast Limburg, as roughly 3,000 miners cycled along Limburg's national highways (Rijkswegen) on workdays.[50] Similarly, researchers reconstructing urban cycling in Rotterdam, determined the cyclists' modal share increased to 82 percent (excluding pedestrians) between 1930 and 1934.[51] Cycling was the favorite travel mode in Dutch industrial areas.

The low cost made cycling an attractive alternative to public transit, trams, and trains. Not just in the Netherlands. Around the world, cities' high tram and train fares motivated many men to revert to cycling, especially when household budgets were under heavy strain due to wage cuts, price hikes, and an unstable job market. *Cycling Cities* research shows that social-democratic city councils favored public transit over cycling. Councils and public-transit operators, however, were forced to realize that the fares did not suit working-class budgets. In response, low-income workers began cycling to work. In Hungary's capital Budapest, when fares were raised during the depression era, city authorities complained about bicycles luring passengers off their preferred urban mode of transport: trams. In Copenhagen too, social-democratic and pro public-transit authorities observed a shift from trams to bicycles.[52] As in other European cities, Dutch workers made a similar affordability trade-off.

Cycling made financial sense. In 1934, public transit engineer Piet Bakker Schut, father of Philips engineer Frits Bakker Schut, calculated in professional

48 Adri Albert de la Bruhèze, "Enschede: An Experiment in Cycling" in *Cycling Cities: The European Experience*, ed. Ruth Oldenziel, et al. (Eindhoven: Foundation for the History of Technology, 2016): 41-51, here 42-43.

49 Manuel Stoffers, "Fietsen in de Nederlandse bergen: Revisited," *SCHL Jaarboek* 61 (2016): 35-42; Frank Veraart and Manuel Stoffers, "Southeast Limburg: Cycling Goes Downhill," in *Cycling Cities: The European Experience*, ed. Oldenziel, et al., 65-75, there 66-68.

50 Eric Berkers, "Fietscultuur en -beleid in Maastricht en Parkstad in historisch perspectief" (Eindhoven: Foundation for the History of Technology, 2017), 9, 15.

51 Eric Berkers et al., *Cycling Cities: The Rotterdam Experience* (Eindhoven: Foundation for the History of Technology, 2019), 20-21.

52 Martin Emanuel, "Copenhagen: Branding the Cycling City," in *Cycling Cities: The European Experience*, ed. Oldenziel, et al., 77-87, here 78, 80; Katalin Tóth, "Budapest: Reviving the Bicycle Lifestyle," in *Cycling Cities: The European Experience*, ed. Oldenziel, et al., 161-171, here 164.

Figure 3: Philips workers cycling over the Woensel railway crossing after their shift. Eindhoven, ca. 1930 (Source: Philips Company Archives)

road-engineering journal *Wegen*, that cycling cost half a cent per kilometer, versus 4 cents a kilometer per bus or tram passenger, concluding that for workers, "The bicycle is by far the cheapest transport mode compared to ... public transit."[53] Even though employers paid travel allowances for bus or rail transport, workers often chose to cycle long distances to save money, especially during the summer months with longer days and milder temperatures. Berkers cites the memoirs of a miner's son, whose father reverted to cycling during summer to "save the tram fare."[54] In another publication on Rotterdam, Berkers et al. discern a similar practice among workers from the city and surrounding villages Barendrecht, IJsselmonde, Rhoon, and Ridderkerk.[55] At the time, Rotterdam-based socialist engineer Johan Nieuwenhuis, examining the relationship between public transit fares and the 1936 bicycle boom, concluded that during the economic downturn, for a growing group of workers and especially unemployed people, a cost-benefit

53 Pieter Bakker Schut, "Het tram-autobus-vraagstuk in verband met de stadsuitbreiding," *Wegen* 10, no. 5 (1934): 116-117. A word of thanks to Henk-Jan Dekker for sharing Bakker Schut's article.

54 "Fietscultuur en -beleid in Maastricht en Parkstad in historisch perspectief," 12-19; Harry Caubo, "Mijn vader, de mijnwerker," Heerlen Vertelt, accessed June 16, 2021, http://www.heerlenvertelt.nl/2016/06/mijn-vader-de-mijnwerker/; Langeweg, *Mijnbouw en arbeidsmarkt*, 66.

55 Berkers et al., *Cycling Cities: The Rotterdam Experience*, 20-22.

analysis favored the bicycle. For Rotterdam's jobseekers (Netherlands' highest unemployment rate of 6.35 percent), the bicycle offered a low-cost mobility alternative to trams and buses, even more so because cyclists were exempt from the relatively high bicycle tax the Dutch government imposed from 1924 to 1941. A one-time investment in a bicycle provided jobseekers flexibility, essential since they relied on temporary jobs scattered across the city or had to seek employment across the wider region.[56]

The Rotterdam socialist engineer also identified flexibility as an important factor in choosing bicycles over public transit. The flexibility that bicycles offered was key for jobseekers, since they had to go wherever local employment officers sent them to work. For them, bicycles were life savers, enabling low-income workers and jobseekers to travel wherever, whenever, cheaply.[57] Today's mobility poverty scholarship centers around the idea of car dependency in expanding people's action radius and accessing key activity locations.[58] In the interwar period, however, the bicycle fulfilled this role for low-income workers. In a period with little alternatives, workers relied on their bicycle rather than on the more expensive train or tram to land jobs and make ends meet.

The Dutch government acknowledged the importance of bicycles for the unemployed men who performed physically demanding, unskilled labor in government work projects like land reclamation in peatlands, afforestation near cities, and road construction. In a 1938 report on the benefits of public works projects for solving unemployment, state inspector J.Th. Westhoff noted that these projects were often in remote places, far removed from any transit connections. Workers had no choice but to accept the work or their unemployment benefit would be stopped, he believed. In addition to a meager weekly wage of between 14 and 17.50 guilders (1939 wage levels), these men were exempt from paying the annual bicycle tax of 2.50 guilders and received cycling allowances. During the cold and wet winter months, workers received 50 cents a week for a daily round trip of thirty kilometers. Some employers provided bus transport for longer distances. In summer, the maximum cycling distance was extended to a round trip of 40 kilometers a day. Workers then received a weekly allowance of 50 cents (10-15 kilometers) or 1 guilder (15-20 kilometers). These allowances, distributed by (local)

56 *Cycling Cities: The Rotterdam Experience*, 20-22; Johan Nieuwenhuis, "Het secundaire vervoer in crisistijd," *De Ingenieur* 51, no. 26 (1936): V35-36.

57 "Het secundaire vervoer in crisistijd," V35-36.

58 Tobias Kuttler and Massimo Moraglio, *Re-Thinking Mobility Poverty: Understanding Users' Geographies, Backgrounds and Aptitudes* (London/New York: Routledge, 2020).

authorities and labor district committees, were expected to cover the cost of bicycle repair and maintenance, thus enabling workers to access remote work locations.[59]

European planners and decision-makers also noted the bicycle's popularity among workers. Though it was reason for authorities to curb the proliferation of "unruly" cyclists in the streets, others were motivated to facilitate cycling commuters. Historian Henk-Jan Dekker highlights in *Cycling Pathways* that every level of government across the Netherlands came to realize how industrial and land workers relied on bicycles for their daily commute. Therefore, labor politicians successfully pleaded for the construction of urban and interlocal bicycle paths—facilitating cyclists, while also making way for cars.[60]

Workers found traveling by bus also an attractive alternative during this period. The first collective taxi and bus services were often private initiatives, organized communally. Paratransit taxi and bus services complemented fixed-route public transit, offering passengers flexible mobility on demand without fixed routes and timetables, providing door-to-door mobility or along loosely defined routes—were yet another mobility alternative that rapidly became popular among workers. According to scholars, these paratransit services were operated by profit-seeking entrepreneurs in the interwar period. Worker communities also operated such bus services.

In Dutch industrial regions like Twente, most tramways were taken over by bus services, bringing workers to and from work. Buses had become a popular, if not essential mobility alternative for many. Small profit-seeking entrepreneurs initially took the initiative, offering bus fares at half the price of rail tickets. Buses, moreover, were more flexible in adapting routes to popular demand than rail-based modes. Railway and tramway companies like NS (ATO), LTM (Limburg), RET (Rotterdam), and TET (Enschede), also began exploiting bus services. As forerunner, Twente's tramway company TET set up bus services already in 1923, steadily replacing trams with

59 "Van het front der werklozen: Een mooi succes voor het Districtscomité voor Werklozenzorg," *De Fabrieksarbeider* 23, no. 50 (1934): 3; J.Th. Westhoff, *De directe mogelijkheden der werkverschaffing bij de werkloosheidsbestrijding* (Zwolle: J.J. Tijl, 1938), 278-279. The national government assigned projects, working hours, and wages, but when the number of unemployed grew during the depression (from 150,000 in 1930, peaking around 600,000 in 1935), government work projects expanded. In Limburg, jobseekers worked on roads and flyovers, in Rotterdam they created the park Kralingse Bos, and in Twente they excavated a canal connecting textile towns Almelo, Hengelo, and Enschede to the main national waterways.

60 Henk-Jan Dekker, *Cycling Pathways: The Politics and Governance of Dutch Cycling Infrastructure, 1920-2020* (Amsterdam: AUP, 2021), 84-85, 106, 118-119.

motorized buses in the following years, transporting textile workers by bus instead of tram. By 1939, already a third of passenger road transport was by bus. Public transit trip numbers remained constant over the 1930s, but the bus share increased from 5 to 31 percent—according to Gijs Mom, indicating the substitution of trams as well an expansion of buses.[61]

Inadequate rail transport motivated profit-seeking entrepreneurs across the globe to set up small scale bus and taxi services. Historians present these grassroots services as business models, ways for local entrepreneurs to make a living from transporting workers and other passengers, adjusted to passengers' mobility needs.[62] Workers and their employers developed similar services, as we will see. For these social actors, this was not an attempt to benefit from transporting passengers. It was a collaborative coping strategy to overcome the transport mismatch between home and workplace.

The bus services were small scale, local endeavors, with company and community owned buses filling a void in the larger, state-subsidized transport system. In a few cases, local authorities organized buses for work crews. If government work projects during the economic depression were not accessible by bicycle or public transit, regional labor authorities organized buses or lorries for workers or housed them in camps.[63] Sometimes employers took the initiative. In Eindhoven, Philips Social Affairs set up small-scale company bus services for employees from surrounding villages, thereby mobilizing workers who lacked affordable (rail-based) alternatives. The *Meierijsche Courant* notified readers in February 1924, that the company had started up a private bus service between Weert and Eindhoven early in the morning and late at night. Initially, Philips signed contracts with regional bus companies. According to one such contract with V.V. Van der Meulen Ansems in Eindhoven, the bus company received a fixed tariff of 1.2 cents per kilometer per seat from Philips: it was contractually obliged to transport 19 workers every working day from

61 Ruud Filarski, "De coördinatiecrisis," in *Van transport naar mobiliteit: De mobiliteitsexplosie (1895-2005)*, ed. Gijs Mom and Ruud Filarski (Zutphen: Walburgpers, 2008), 204-210, 229-232; Gijs Mom, "Clashes of Cultures: Road vs. Rail in the North Atlantic World during the Interwar Coordination Crisis," *The Organization of Transport: A History of Users, Industry, and Public History*, ed. Massimo Moraglio and Christopher Kopper (New York: Routledge, 2015): 18-37, here 9-10.

62 Filarski, "De coördinatiecrisis," 203-235; Mathieu Flonneau, "Collective Taxis in 1930s Paris: A Contribution to an Archeology of 'Uberization'," *The Journal of Transport History* 39, no. 1 (2018): 12-24; Mom, "Clashes of Cultures," 18-37.

63 "Van het front der werklozen," 3; Westhoff, *De directe mogelijkheden der werkverschaffing bij de werkloosheidsbestrijding* 278-279.

Figure 4: Women workers boarding one of Philips' first company buses. Eindhoven, ca. 1924 (Source: Philips Company Archives)

Tilburg to Eindhoven along "the shortest routes" and "officially stipulated by the province Noord Brabant." Apparently, it did not suffice. Several years later, Philips established its own bus services (fig. 4) servicing other places like Uden and Bommelerwaard in the province Gelderland, according to *Eindhovens Dagblad* in 1928.[64]

Philips wrote to Eindhoven city to expand these company bus services to the Belgian villages Overpelt, Neerpelt, and Bree (60 km from Eindhoven), approved by the council, which supervised state-imposed passenger transport safety regulations.[65] In the *Meierijsche Courant* (February 1929), Philips notified (potential) workers about new services and routes—bus stops at Poort Glasfabriek A, Stratumseind, and Bloemenplein, as well as departure times, boarding procedures, and fares (30 cents a week).[66] According to Philips housing engineer Bakker Schut, the Philips buses were not only

64 PCA, Archive no. 642.5, Inv. no. 1-1929.02, Internal Minutes 25 February 1928, Contract 28 March 1928, 1; "Philipswagen Weert-Eindhoven," *Meierijsche Courant*, 16 February 1924 ; "Philips eigen autobusdienst," *Eindhovens Dagblad*, 21 September 1928, 6; "Philipspersoneel uit Gelderland," *Eindhovens Dagblad*, 28 February 1930, 2.

65 PCA, Archive no. 642.5, Inv. no. 1-1929.02, Correspondence 5 November 1927.

66 "Philips' autobusdienst; 8 lijnen; 15 bussen," *Meierijsche Courant*, 18 February 1929.

cheaper than public buses but offered workers a faster mobility option for the price of cycling. He calculated that cycling cost 90 cents at an average speed of 15 kilometers per hour, the same as Philips buses (90 cents per kilometer), though with the bonus of traveling 20 kilometers per hour faster.[67]

Philips bus services were popular with employees. Not only workers living in the surrounding countryside, but also on the outskirts of Eindhoven requested Philips bus services—suggesting this was a company initiative but workers had their say. In November 1929, workers living in the Bloemenkwartier in Stratum petitioned the Philips board. Up for consideration was an extra bus service to the factory on account of the more than an hour's walk there, which "was very unpleasant in bad weather"[68] Thirty-seven employees signed the petition with their name, employee number, and department, ranging from engineers in the research and development Nat-Lab, to women working on the glassworks assembly line.[69] Philips workers thus negotiated their commute through such written requests, even though the buses were essentially a company instrument to mobilize rural labor forces.

Limburg's mining workers, in contrast, took the initiative to run bus services. Several miners, fed up with rail transport, set up their own local bus services because their routes and timetables were more flexible to the needs of the generally well-paid miners. These first miner buses picked up co-workers for a break-even fare in self-purchased lorries refurbished by local wheelwrights. In 1938, union magazine *De Mijnwerker* recalled that such a bus was like a huge "box" with rough benches and walls barely protected from the wind and rain, "nailed onto an old leggy Ford truck by a local wagonmaker."[70] When the first autobus appeared in Southeast Limburg in 1924, miners living outside the mining district were delighted. Buses proved the "ideal transport mode" because "they picked up workers from their door and put an end to the tiring walk and demoralizing wait" for trains and trams, *Algemeen Handelsblad* and *Nieuwe Venlosche Courant* noted. Miners' commuter time was significantly reduced and the bus cost just a fraction of train and tram fares. The miners' union magazine praised the introduction of these buses: the unknown miner who came up with

67 Bakker Schut, *Industrie en woningbouw*, 326-329. He included estimated costs for bicycle and clothing's wear and tear, and the cost of time lost en route.

68 PCA, Archive no. 642.5, Inv. no. 1-1929.02, Correspondence 15 November 1929, 1-3, 17 December 1929, 4 January 1930.

69 Ibidem. Workers had to keep on walking or cycling to work: management could not honor the request, because they did not have sufficient buses available.

70 "Het vervoervraagstuk der mijnwerkers," 363-364.

the idea to transport miners by autobus "deserved a statue somewhere in Limburg," the editors thought in 1938. Gradually, local entrepreneurs noted the potential profit from bus exploitation and stepped in, even former miners. Between 1924 and 1925, almost forty bus companies were set up in Southeast-Limburg alone.[71]

Autobus services were in demand with miners living 20 to 30 kilometers from Heerlen's mine. Mining companies did not follow Philips' example of company buses. They did acknowledge the advantages of these services for improved productivity and lower absenteeism among miners in remote villages, no longer forced to travel by unreliable trains and trams or make long, exhausting bicycle rides over Limburg's hilly landscape. Traditionally, mining companies were directly involved in rail expansion in Limburg—sitting on national committees and making investments. They were only indirectly involved in buses, according to J.M.A.G. Gillissen's dissertation *De invloed van de Staatsmijnen in Limburg op de ontwikkeling van het locale personenvervoer en op de woningbouw in de mijnstreek tot negentien honderd en veertig* (The Influence of the Limburg State Mines on the Development of Local Passenger Transport and Housing in the Mining Region until 1940 (1962)). Mining managers supported bus transport development by guaranteeing local bus operators a minimum number of passengers, and negotiating fares and routes with operators in consultation with miners' union representatives.[72] Unlike Philips company-owned bus services, this more aloof attitude of the mines towards worker buses, led to serious problems by the late 1930s, when government regulations raised cost barriers for miners as the following chapter shows.

71 "Verkeersstaking in Z.-Limburg: Weinig gewaardeerde ordening," *Algemeen Handelsblad*, 15 May 1938, 4; "De vervoerkosten van en naar de mijnen," *Nieuwe Venlosche Courant*, 4 June 1938, 8; "Het vervoervraagstuk der mijnwerkers," 363-364; Langeweg, *Mijnbouw en arbeidsmarkt*, 66-67.

72 RHCL, Archive no. 17.05H, Oranje-Nassau Mijnen: Vervoer, 1903-1978, Inv. no. 46, Vervoer mijnwerkers per autobus, Correspondence no. 604, no. 1028, no 1752; J.M.A.G. Gillissen, *De invloed van de Staatsmijnen in Limburg op de ontwikkeling van het locale personenvervoer en op de woningbouw in de mijnstreek tot negentien honderd en veertig*. (PhD Dissertation, Sociaal Historisch Centrum Limburg, 1962), 14, 33-34; Langeweg, *Mijnbouw en arbeidsmarkt*, 67. Fares were between 0.25 and 0.70 guilders for a return ticket and 0.10 to 0.40 for one-way. The 1931 timetable of new P.L. Brouns bus service from Stein to Sittard, published in *Limburgsch Dagblad*, shows that bus stops were strategically located in mining town centers like at Café Smeets and Café Canen in Stein, State Mine Maurits, near municipal bicycle sheds ("Rijwielloods"), and Geleen market square. "Aankondiging Dienstregeling," *Limburgsch Dagblad*, 3 January 1931, 3; "De vervoerkosten van en naar de mijnen," 8.

Conclusion

Against the backdrop of interwar industrialization, labor market expansion, and the economic downturn, rail-based transit fell short in meeting workers' and industries' mobility needs. The more rigid and expensive rail-based system no longer offered solace, and bottom-up mobility alternatives emerged. Bicycles and buses—bare-bone mobilities as they required little additional infrastructural investment—became popular options to get to factories, mines, and government projects. The locus of control to decide how and why to move, came to rest more with workers and local entrepreneurs rather than large railway companies and the state—except for the workers assigned to government work projects. Individually, self-governing workers appropriated and repurposed cheap bicycles for commuting, expanding their action radius.

The first worker bus services represented another bottom-up response to transport mismatch: workers and employers co-creating small scale, shared paratransit services. The future was uncertain, however, when the state stepped in to regulate the sector the Ministry of Public Works raised cost barriers for low-income passengers by imposing strict bus regulations. What did the state's intervention mean in terms of workers' (im)mobility and job access? And how did workers respond?

2 Protesting Bus Regulations during the Depression, 1926-1938

Sunday May 8, 1938, shortly after mass, miners in Southeast Limburg gathered in boisterous protest meetings in response to state-imposed bus fare hikes. In an effort to impose order on interwar road transport, the Dutch government acted vigorously against paratransit bus operators. Tightened rules for passenger buses spiked fares. Employers largely remained indifferent. But blue-collar workers, outraged by the higher fares and exclusion from decision-making, took matters into their own hands. In various mining towns, hundreds of miners gathered. Local union leaders called for a bus boycott, urging miners to cycle to work the next morning.[1] In the following days, people witnessed "cycling demonstrations" of miners along Limburg's roads, and dozens of empty buses arriving at the mine shafts, "escorted by long strings of cycling miners."[2] Regional newspaper *De Nieuwe Venlosche Courant* estimated that between 2,000 and 4,000 miners participated in the bus boycott. This was an astonishing number, union magazine *De Mijnwerker* thought, given that Dutch miners rarely engaged in strikes over pay or working conditions.[3] Why were these unionized miners so agitated by these price hikes? Why would miners earning relatively good wages rather cycle up to thirty kilometers a day over hilly terrain than pay higher fares?

Mobility historian Gijs Mom and others have examined the interwar government coordination of road and rail passenger transport. They indicate that increasing state regulation led to soaring bus fares and dwindling numbers of bus operators, but do not address whether this raised mobility

1 "Tarieven mijnwerkersvervoer," *Limburger Koerier*, 10 May 1938, 5; "Mijnwerkers boycotten autobussen: Felle twist in Zuid Limburg," *De Telegraaf*, 17 May 1938, 4; "Actie tegen verhoogde vervoerprijzen: Mijnwerkers boycotten autobusvervoer," *Leeuwarder Courant*, 17 May 1938, 10.

2 "Autobussen werden te duur," *Utrechts Volksblad: Sociaal-Democratisch Dagblad*, 17 May 1938, 1.

3 "Rondom de autobus-kwestie," *De Mijnwerker: Orgaan van de Ned. R.K. Mijnwerkersbond*, 28 May 1938, 170; "De vervoerkosten van en naar de mijnen," *Nieuwe Venlosche Courant*, 4 June 1938, 8.

Bek, Patrick, *No Bicycle, No Bus, No Job: The Making of Workers' Mobility in the Netherlands, 1920-1990*. Taylor & Francis Group 2022
DOI: 10.5117/9789463723183_CH02

barriers for workers.[4] "Protesting Bus Regulations" reveals that state-imposed regulations raised cost barriers for workers considerably. While employers did not react, workers openly protested this course of events. Despite labor history's longer tradition of studying protests as an important form of subaltern agency, it has not dealt extensively with mobility as a domain of protest.

Labor historians describe protests as an important element of workers' action repertoire. In *Workers of the World*, Marcel van der Linden summarizes the boycott scholarship as follows: a protest was "more or less coordinated action by a group of workers (and, perhaps allies) to attain a specific objective, which they would be unable to achieve individually within the same time frame with the means available to them." Protests were common for improving labor relations, against allegedly unjustified price increases of foodstuff, clothing, and housing due to market changes, taxes, and state-imposed regulations. Boycotts arose spontaneously or were organized by political organizations and unions (looking to strengthen their position), and not seldom served a symbolic purpose, expressing a "sense of unity" or the sentiment "we've had enough."[5] Especially for individuals and collectives with little access to power and little room to maneuver, protests are important strategies to influence the course of developments.

The issue of workers' mobility and commuting also became a potential domain for political action. Historian Alf Stadler, reflecting on 1940s bus boycotts in Johannesburg, South Africa, claimed that a seemingly marginal price increase of a penny on the bus fare profoundly raised cost barriers for poor black workers living in urban communities far from workplaces. Bus boycotts illustrated occasions when workers intervened in the direction of

4 Ruud Filarski, "De coördinatiecrisis," *Van transport naar mobiliteit: De mobiliteitsexplosie (1895-2005)*, ed. Gijs Mom and Ruud Filarski (Zutphen: Walburgpers, 2008): 203-235; "The Emergence of the Bus Industry: Dutch Transport Policy During the Interwar Years," *Transfers* 1, no. 2 (2011): 61-82; Gijs Mom, "Clashes of Cultures: Road vs. Rail in the North Atlantic World during the Interwar Coordination Crisis," *The Organization of Transport: A History of Users, Industry, and Public History*, ed. Massimo Moraglio and Christopher Kopper (New York: Routledge, 2015): 18-37.

5 Marcel van der Linden, *Workers of the World: Essays Toward a Global Labor History* (Leiden/Boston: Brill, 2011), 11, 209, 212-213, 217-218. The history of technology describes protest as an important user strategy to oppose and/or develop technological (mobility) systems. Building on the work of labor historians, historians of technology Van der Vleuten, Oldenziel, and Davids synthesize scholarship on user activism. Users of technological artefacts and systems applied a varied repertoire of activist strategies: communicating core values via printed media; direct actions, sit-ins, and sabotage; tweaking existing systems for their own uses; and developing alternatives. Mobility systems have been the object of such strategies when confronted with high costs or mobility mismatches. Erik van Vleuten, Ruth Oldenziel, and Mila Davids, *Engineering the Future, Understanding the Past: A Social History of Technology* (Amsterdam: Amsterdam University Press, 2017), 133-138.

their lives.[6] For 1970s Italy, Bruno Ramirez shows how amid inflation and widespread unemployment, hundreds of workers from Pinerolo, employed in Turin fifty kilometers northeast, refused to comply with spiking bus fares by setting up signs near bus terminals, saying "Refuse the fare increase!" and issued substitute bus tickets for the old price.[7] In both cases, the workers' protests were successful, as the transport companies reinstated the old fares. The dominant state-supported transport companies, however, upheld the power to reinstate fares and adjust regulations. The question remains to what extent workers could influence decision-making processes that affected them getting to work.

As the Limburg strike indicates, "Protesting Bus Regulations" highlights how workers and their unions shaped their everyday mobility by appropriating options, fending for themselves, and contesting state-imposed mobility barriers with political protest. It reveals the underlying state politics of control over road transport that led to spiked bus fares in the Netherlands. It illustrates the social impact on miners through a case study of Vaals, a small mining town in Southeast Limburg, highlighting the miners' political agency in responding to increased cost barriers. Although most miners' union archives were destroyed during the Second World War, including first-hand accounts of the 1938 strike, local journalists and union magazines covered the boycott extensively. They provide insight into how and why working-class miners actively intervened in their mobility and the direction of their lives.

2.1 State Regulation of Passenger Buses

For workers and industrial employers facing a transport mismatch, small scale bus services filled a void in the Dutch transport system. These ventures also challenged the existence of the incumbent (rail) regime.

6 Alfred Stadler, "A Long Way to Walk: Bus Boycotts in Alexandra, 1940-1945," paper presented at the *African Studies Seminar*, University of the Witwatersrand, Johannesburg, South Africa (1979), 2, 6, 13-14, 18. In *Africa South* (1957), anti-Apartheid activist Ruth First characterized the 1943 boycotters as "an army of protesters, voting with their feet," as a procession of 15,000 workers walked more than 15 km from their townships to work. Similar to Limburg, at Johannesburg's Great Bus Boycott of 1957, First observed "great lumbering green buses of the largest transport organization for Africans in the country travelled empty along route ... for five and six hours every day endless streams of walkers filled the pavements." See Ruth First, "The Bus Boycott," *Africa South* (July-September 1957): 55-64, here 55, 59. Bicycles proved to be a vital mobility alternative to black workers: Njogu Morgan, *Cycling Cities: The Johannesburg Experience* (Eindhoven: Foundation for the History of Technology, 2019), 36-37.

7 Bruno Ramirez, "The Working-Class Struggle Against the Crisis: Self-Reduction of Prices in Italy," *Radical America* 10, no. 4 (1975): 143-150, here 144-145.

Dutch Railways—also shareholder in many regional tramways—saw the proliferation of bus companies as a major threat to their revenues. Unlike rail-based transport, bus owners had lower running costs and as a new service were not yet bound by legal regulations regarding routes, fares, and safety standards. In an unregulated market, bus entrepreneurs could operate without licenses wherever they expected to make a profit. For passengers, bus travel meant direct connections and reduced travel time, hence its general popularity. Bus companies mushroomed across Europe during the 1920s. They offered services alongside rail- and tramways for half the price, and rail companies saw passenger numbers dwindle.[8] Wherever passenger buses proliferated, Ruud Filarski notes that governments faced a dilemma: "should they intervene in the market to establish a level playing field for fair competition between the buses and rail transport, should they protect the loss-making railways, or should they take a laissez-faire approach?"[9] The common answer by governments, siding with rail companies struggling to maintain their market share, was to regulate buses.

State regulation of the popular buses to protect railroad interests happened in the Netherlands and beyond. Soon after Dutch bus operators crossed the Belgian border to provide their services in the densely populated northern provinces, the Belgian government introduced a licensing system in 1923, protecting their national railroads' interests. The German Road Transport Services Act (1928) and British Road Traffic Act (1930) introduced similar licensing systems for bus operators, though they led to more rigorous bus regulations in the U.K.[10] Historian Reg Davies shows in his thesis on passenger transport in interwar Britain, that Southern Railways, confronted with dwindling passenger numbers, argued strongly for state control over the bus industry in the 1920s. The result was the 1930 Act that aimed to rationalize passenger transport by eliminating duplication of services by operators of the same or different modes of transport. This was often to the detriment of bus operators. Bus fares were generally lower than rail fares, but bus companies were forced via these licensing instruments to operate for the same prices.[11]

8 J.M. Fuchs, *Onderweg: Een eeuw personen- en goederenvervoer in Nederland* (Amsterdam: KNVTO, 1981), 106-109; Bert Nijkamp, "Herinneringen aan het openbaar autobusvervoer in Zuid-Limburg: De jaren na de Tweede Wereldoorlog," *Het Land van Herle* 52 (2002): 1-13, here 3-9; Luc Wolters, *Limburgse tijdreis met bus en tram: 90 jaar gemeentelijke activiteit in vervoermobiliteit* (Maastricht/Sittard: Gebaltram, 2010), 42-43.

9 Filarski, "The Emergence of the Bus Industry," 61-82.

10 Mom, "Clashes of Cultures," 11-12.

11 Reg Davies, *Public Passenger Transport in Inter-war Britain: The Southern Railway's Response to Bus Competition, 1923-39* (PhD Dissertation, University of York, 2014), 125, 130, 135-136.

The state claimed to "tame" what it characterized as the unbridled proliferation of unsafe and unfair passenger transport. In fact, it singled out bus companies and restricted workers' mobility by bus. In response to the competition with rail services, the Dutch national government regulated the bus services so popular with low-income people in the mid-1920s. Historians Mom and Filarski, in their comprehensive history of transport in the Netherlands, recall that in 1926, at Dutch Railways' request, the national government intervened by adjusting the 1880 Public Means of Transport Act (Wet Betreffende de Openbare Middelen van Vervoer). Bus companies now had to acquire licenses, and as a result, many were prohibited from operating. Local bus company NEDAM, that transported 125,000 passengers a year in Limburg's mining region, was banned. Scholars show how the debate lingered on until 1937, when the government intervened with another act and the Ministry of Public Works Agency (Waterstaat) established the Commission for Passenger Transport Licenses (Commissie Vergunningen Personenvervoer, CVP, later Commissie Vervoersvergunningen, 1959), headed by Professor Ir. Hendrik van Breen, a civil engineer with extensive experience in the Dutch colonies, president of the General Dutch Traffic Federation (Algemeene Nederlandsche Verkeersfederatie), and sympathetic with Dutch Railways. This Commission redistributed public transit concession areas, and set the standards for safety requirements, personnel, and fares—which formed the basis for the post-World War II public transit system. After a semi-official implementation in 1938, the Commission fully implemented the Motorized Passenger Transport Act in 1939 (Wet Autovervoer Personen, WAP): all road passenger transport companies had to register with the CVP and request a permit for passenger services. Unlike the provincial bureaus established in 1926, the CVP subjected transport companies to more strictly demarcated concession areas, leading to a drastic reduction in bus services and further concentration of transport companies. The Netherlands counted 520 small bus companies (ten or less employees) in 1928. That number was reduced to 350 companies granted concessions, not for specific routes but larger so-called transport areas. Bus services like Philips between workers' villages and Eindhoven factories also stopped.[12]

The Ministry, siding with Dutch Railways, set the standards and requirements for passenger transport concessions, then used those standards to disadvantage their rivals (bus companies). According to Mom, this regulation stemmed from a desire for order among technocratic civil engineers and

12 Philips Company Archives, Archive no. 642.5, Personeelsvervoer: Ned-NV VIPRE, Inv. no. 882, Correspondence 31 December 1948, 1; Filarski, "De coördinatiecrisis," 204-210, 229-232; Fuchs, *Onderweg*, 110-112, 188-190; Wolters, *Limburgse tijdsreis*, 50-51.

decision-makers, even though it did not serve the public good. Governing (un)fair competition between road and rail transport was their guiding principle—what defined fair or unfair was up to the establishment. This division of road transport in desirable and undesirable modes of transport helped the Ministry of Public Works to make passenger transport governable and set the norm which bus operators had to assimilate.[13]

The state may have sided with the vested interests of the railroads for economic reasons. Yet, the policy intervention also negatively impacted low-income people's mobility and created new problems. First, smaller bus companies were virtually unable to obtain licenses and had to shut down. Second, the new legal requirements—safety standards, insurance, trained personnel and so on, meant higher operational costs for licensed buses and thus higher fares for passengers. With the number of passenger buses on Dutch roads dwindling and bus fares spiking, state regulations generated mobility barriers for workers (or mobility scarcity, in scholars' terms). Restricting people's movements by bus shifted the problem of those in power (the state and Dutch Railways) onto working-class passengers, who relied on the buses to get to work.

Bus entrepreneurs and passengers responded with several tactics. Across the country, bus owners organized protest meetings and launched court appeals. Passengers covertly supported local bus companies, by pretending they were on group trips. These efforts aroused public agitation, but had little effect.[14] The state threw in its full weight: clandestine drivers were arrested, bus materials confiscated, and entrepreneurs running "illegal" passenger services were convicted. "An end to the Autobus War?" newspaper *De Telegraaf* asked in 1935: not for long.[15] In 1938, State Traffic Inspectors' newly imposed bus regulations again led to price hikes and the "autobus war" flared up once more.

The new status quo met fierce resistance from Limburg's miners and their unions. Facing increased cost barriers, miners in their thousands

13 Mom, "Clashes of Cultures," 18-21, 28. For discussion of "categorical bordering" Saurabh Arora et al., "Control, Care, and Conviviality in the Politics of Technology for Sustainability," *Sustainability: Science, Practice and Policy* 16, no. 1 (2020): 247-262.

14 Greet De Block, "Urbanizing the Countryside: Rails, Workers and Commuting in South-West Flanders, Belgium, 1830-1930," in *Cultural Histories of Sociabilities, Spaces and Mobilities*, ed. Colin Divall (London: Routledge, 2015): 53-66, here 65; Paolo Capuzzo, "Transport and (sub)urban development: Between politics and technology: transport as a factor of mass suburbanization in Europe, 1890-1939," *Suburbanizing the Masses: Public Transport and Urban Development in Historical Perspective*, ed. Colin Divall and Winstan Bond (London/New York: Routledge, 2003): 23-48, here 33; Gijs Mom and Ruud Filarski, *Van Transport naar Mobiliteit: De Mobiliteitsexplosie 1895-2005* (Zutphen: Walburgpers, 2008), 33.

15 "Autobus-oorlog ten einde?," *De Telegraaf*, 17 October 1935, 3.

exercised collective consumer power by boycotting regional buses and reverting to cycling to work. Miners militated against the technocratic state's radical monopoly driven by ambitions to control road transport. A closer examination of this boycott reveals the everyday impact on workers and their communities. It proves additional evidence—and in more detail—how bicycles were an essential mobility alternative for low-income workers.

2.2 Workers Respond with a Miner Bus Boycott

Workers' preference for bicycles over public buses or trams is what historians characterize as a form of worker protest against high fares. Local (social-democrat) authorities across Europe were not happy with this trend, as they envisioned mobilizing the working-classes via rail-based public transit, as we saw. In Hungary, Budapest's public authorities, confronted with loss-making trams, even framed people's shift to bicycles as deliberate "sabotage," Katalin Tóth argues.[16] The Dutch miners' bus boycott in 1938 and their collective switch to bicycles were political actions too, albeit a more organized and visible expression of workers' resentment towards transport company and government policies. Miners were not just agitated about the price hikes. They also were disgruntled with the Ministry of Public Works' entire decision-making process.

Cycling was a form of resistance against the mobility regime. For Limburg's protesting miners, the bicycle was their vehicle of choice, enabling them to enforce their political statement without forfeiting their ability to access jobs and make a living. In March 1938, regional newspapers and union magazine *De Mijnwerker* reported imminent price hikes in May 1938. Limburg State Inspector for Traffic (Rijksinspecteur van het Verkeer) Max Meijer's announcement did not immediately alarm the public. Miners were convinced their unions would reach an agreement with the State Inspector and Minister of Public Works. In Peys, several miners gathered at a meeting chaired by Wiel Mulders, vice-president of the Catholic Miners Union (Nederlandse Katholieke Mijnwerkersbond, NKMB, and member of the Catholic People's Party (KVP). Miners voiced their concerns. They felt that the new concession requirements under the Autobus Act did not

16 Ruth Oldenziel and Adri Albert de la Bruhèze, "Contested Spaces: Bicycle Lanes in Urban Europe, 1900-1995," *Transfers* 1, no. 2 (2011): 29-49; Katalin Tóth, "Budapest: Reviving the Bicycle Lifestyle," *Cycling Cities: The European Experience*, ed. Ruth Oldenziel et al. (Eindhoven: Foundation for the History of Technology, 2016): 161-171, here 164.

justify the price increase—and crucially, as major stakeholder, had been ignored by the State Inspector or the Ministry. Mulder found that miners should have had a say. And, he added, bus exploiters should also actively cooperate not just with the state, but also with miner communities to agree on reasonable fares and safeguard miners' access to bus services. If the government and the bus companies failed to do so, Mulder argued, miners should all cycle to work, and feel confident about their political leverage. Within a few days, the State Inspector and bus companies "will be whistling a different tune," he predicted.[17]

Miner resistance was not solely about the fares. Their protest also questioned the legitimacy of the decision-making process: a "silent agreement" between bus companies, the State Inspector and the Autobus Commission headed by Van Breen, a fervent fan of the railways. Van Breen's Commission had consulted several bus companies about setting new fares in September 1937 already, but no-one had consulted the miners or their representing organizations beforehand. Miners and their unions were excluded from decisions that profoundly impacted their lives, according to the regional press and *De Mijnwerker*. They wrote that decision-makers at the Dutch government in faraway The Hague were insensitive to the lived experiences of hard-working miners in the South. What the policymakers determining bus fares had not taken into account, was that higher bus fares would also affect miners' access to everyday mobility. This "wallet issue" as the catholic union called it, affecting their ability to commute to mining sites, would impact miners' household incomes and housing locations.[18] Despite miners' relatively high wages compared to workers in other industries, price hikes were a significant drain on household budgets in an economically unstable time.

The press and union magazines might have inflated the reports and used sensational wording, but other sources confirm the severity and social impact. Correspondence from Vaals town council provides a similar perspective of the miners and their families. The technocratic regulation of interwar passenger buses illustrates how state-imposed regulations—drawing arbitrary boundaries

17 "Bus-vervoer mijnwerkers," *Limburger Koerier*, 22 March 1938, 6; "Nogmaals de autobus-kwestie," *De Mijnwerker: Orgaan van de Ned. R.K. Mijnwerkersbond*, 28 May 1938, 172. Mulders tried to temper the agitated miners, urging them to bear in mind all that had improved since the "traffic anarchy of earlier days." ("verkeersanarchie in vroegere jaren"), referring to the early 1920s. The State Inspection of Traffic was also a first step in keeping a watchful eye on bus exploiters, pricing, routes, and safety requirements.

18 "Fiets contra autobus: Mijnwerkers tegen tarievenpolitiek," *Limburger Koerier*, 14 May 1938, 5; "De hetze van de socialistische bond," *De Mijnwerker: Orgaan van de Ned. R.K. Mijnwerkersbond*, 4 June 1938, 178; "De vervoerkosten van en naar de mijnen," 8.

by setting prices, controlling entries, and regulating conduct—hampered the development of these buses that were a grassroots solution to mobility scarcity. Consequently, workers were forced to choose bus services sanctioned by state officials who seemed indifferent to local people's mobility needs.

The discontinuation of the bus services also impacted local communities economically. Several towns feared that mining companies—contracting bus operators, but mostly not involved in the bus debate—would force their employees to move to company housing locations in the vicinity of the mines. Vaals, a town outside the mining district with a large miner community, depended on local bus services for daily 20 to 30 km commutes to the State Mines in Hoensbroek and Lutterade. The mayor and aldermen expressed their concerns: miner families moving due to increased bus fares, would mean a drain of 160 families from Vaals—70 miners working at Maurits in Lutterade, 90 at Emma in Hoensbroek. The council took up the case for miners and their families, attempting to avert a socioeconomic "catastrophe" for the community of Vaals. The projected exodus of many miner families would have a negative impact on the community, local retailers, and farmers, and ultimately also threaten the local bus companies, already seeing their passenger numbers and therefore profits drop rapidly as miners boycotted buses and cycled to work.[19]

Travel costs took up a big chunk of workers' disposable income. Miners may have earned better wages—on average 5.01 guilders a day in 1938 compared to 3.27 to 3.88 for other workers—but their commuting costs were also higher.[20] Since the mining locations in Southeast Limburg were regionally dispersed—unlike other industrial regions in the Netherlands—travel distances and costs were greater for miners using buses. For miners in Vaals, bus fares to state mines Maurits and Emma increased from 2.50 to 3.50 guilders a week. This meant that miner households previously spending 8.3 percent of their weekly income on bus travel in 1938, saw this rise to 11.3 percent due to the hike in bus fares. This 3 percent might seem like a minor increase. Relatively, however, it meant miners had to pay a third more for commuting than before.[21]

19 Gemeentearchief Vaals (GAV hereafter), Archive no. 2094, Archieven der Gemeente Vaals (1893) 1930-1981 (2000),Verkeer en vervoer: Zorg voor en toezicht op middelen van vervoer, "Correspondentie met de Commissie Vervoersvergunningen, de Rijksverkeersinspectie en het Ministerie van Waterstaat" (1938), Correspondence no. 607 and 847; "Verkeersstaking in Z.-Limburg: Weinig gewaardeerde ordening," *Algemeen Handelsblad*, 15 May 1938, 4.

20 Peter Schrage, Erik Nijhof, and Piet Wielsma, "Inkomensontwikkeling van werkenden en werklozen in Nederland (1913-1939)," *Tijdschrift voor Sociale Geschiedenis* 14 (1989): 347-394.

21 GAV, Archive no. 2094, Inv. 2094, Correspondence no. 705; "Vervoer van mijnwerkers: Bushouders willen prijsverhoging, mijnwerkers gaan per fiets," *Limburger Koerier*, 29 March 1938,

The mining community Vaals expressed their economic reason for cycling to work by protesting. Vaals alderman Huub Hermans, labor party member and active miner, voiced his concerns that the higher travel costs would have a disastrous impact on the community. He argued that in the public interest, affordable mobility options should outweigh government considerations regarding the regulation of unfair competition between bus companies. According to the *Limburger Koerier*, Hermans was amazed that in a country where "the golden freedom was so admired and sung, that same freedom was limited when it came to miners' mobility."[22] His views resonated with Vaals town council. The mayor Hubertus J. Rhoen, backed by local bus owner J.H. Geelen, wrote in protest to the Minister of Public Works, the liberal J.A.M van Buuren. Notwithstanding higher fares, the Vaals miners were now also obliged to buy prepaid bus tickets, with no possibility of restitution for days not travelled due to sickness or holidays, i.e. days they did not earn anything.[23] Not only were miner communities denied a say in the decision-making. State officials also ignored them afterwards, despite the profound impact of the regulations on communities. Until the case was resolved, miners cycled the twenty or thirty kilometers to the mines.

In response, the miners held a bicycle protest. On Sunday, May 8, 1938, protest meetings were organized at venues and pubs across Southeast Limburg, attracting hundreds of local miners. Such meetings were commonly chaired by local representatives of the catholic or socialist miners' unions, and occasionally visited by representatives of bus companies, concession holders of the routes in question. Press reports on these "boisterous meetings" show how the miners living outside the mining districts, who depended on affordable bus transport, were the most agitated. At the pub De Brok in Roermond, dozens of miners from nearby places Herkenbosch, Swalmen, Maasniel, Horn, Herten, and Linne assembled in protest, *Limburger Koerier* reported. Hundreds of miners gathered from Eys, Simpelveld, and Brunssum, who decided that if the fare increase was not reversed, they would boycott

7; "Actie tegen verhoogde vervoerprijzen: Mijnwerkers boycotten autobusvervoer," 10; "Autobussen werden te duur," 1; "De vervoerkosten van en naar de mijnen," 8; Serge Langeweg, "De geographische spreiding van de mijnwerkersbevolking," in *Mijnwerkers in Limburg: Een Sociale Geschiedenis*, ed. Ad Knotter, et al. (Nijmegen: Van Tilt, 2012): 100-138, here 116-117. Own calculations of old fares: (2.5/(5.01*6=30.06)*100=8,3 percent; and new fares: (3.5/(5.01*6=30.06)*100=11.3 percent

22 "Vervoer van mijnwerkers: Bushouders willen prijsverhoging, mijnwerkers gaan per fiets," 7; "Het vervoer der mijnwerkers," *Limburger Koerier*, 14 May 1938, 2.

23 GAV, Archive no. 2094, Correspondence No. 607 and 847; Nationaal Archief, Archive no. 2.16.83 Archief van het Ministerie van Waterstaat: Bureau Autobussen, 1943-1950/Commissie Vergunningen Personenvervoer, 1937-1959, Inv. no. 38, Stukken betreffende besprekingen omtrent mijnwerkersvervoer, 1941-1942, "Correspondentie met College van B&W van de Gemeente Vaals", Mei 1938.

the bus services. At event venue Peters in Susteren, miners—the majority working at Maurits State Mine and facing a bus fare increase from 1.50 to 1.75 guilders a week with no possibility for restitution if they travelled less than six days in a week for whatever reason—met with bus operators. Rejecting the operators' explanation of pressure to comply with the new regulations for fear of losing their concession, the protesters threatened to revert to cycling if this injustice was not undone. In Pey too, over 300 miners gathered, chaired by Jeuken, president of the local Catholic Workers' Association (Rooms-Katholieke Werkliedenvereniging). Mulders of the union NKMB appealed the State Inspector's decision. To make a political statement, the miners in Pey announced they would cycle to work the next day.[24]

On Monday morning, miners boycotted the buses and cycled to work. Miners in Thorn waited at the bus stop as usual, then demanded to travel for the old fares. When the driver did not allow them on board, the miners went home to collect their bicycles and cycled to the mines. In the towns of Echt, Roermond, and Susteren, miners set out in large groups at four in the morning, cycling 20 to 30 km to Maurits Mine in Lutterade, leaving buses on the route largely empty except for the 10 percent of usual commuters. Along with the miners from places near Maurits who normally cycled to work, this resulted in "an invasion of cyclists at the Maurits Monday morning," the *Limburger Koerier* marveled.[25] *Limburgsch Dagblad* noted that, partly because of the fine weather, the miners were in good spirits, perceiving the protest as a pleasurable trip. Especially the younger miners cycled. Others who were unable to cycle because of disability or age, arrived in taxis, omnibus trains ("boemeltreintjes"), or the mining companies' infrequent bus services.[26]

Miners also accused the Ministry and state traffic inspectors of failing to recognize their interests. By means of a boycott and cycling protest, miners took collective action, rallying a critical mass against the establishment. In short, the miner bus boycott was a bottom-up act against the state's bus mobility regulations, which miners and their unions found unjust and

24 "Vervoer van mijnwerkers: Bushouders willen prijsverhoging, mijnwerkers gaan per fiets," 7; "Tarieven voor mijnwerkersbussen: Actie tegen de vastgestelde verhoogingen," *Limburger Koerier*, 29 March 1938, 6; "Tarieven mijnwerkersvervoer," 5.

25 "De verhoogde tarieven voor mijnwerkersvervoer: In Echt en Roermond boycott der bussen," *Limburger Koerier*, 10 May 1938, 6; "Thorn: Autobusvervoer mijnwerkers," *Limburger Koerier*, 12 May 1938, 6; "Susteren: De mijnwerkers gaan per fiets," *Limburger Koerier*, 12 May 1938, 7; "Per fiets!," *Limburger Koerier*, 13 May 1938, 6; "Het vervoer der mijnwerkers," 2; "Verkeersstaking in Z.-Limburg," 4.

26 "Protest van mijnwerkers tegen de nieuwe autobustarieven," *Limburgsch Dagblad*, 10 May 1938, 7; "Het vervoer der mijnwerkers," 2; "Verkeersstaking in Z.-Limburg," 4.

undemocratic. In doing so, these workers were making mobility collective forming or reinforcing the social practice of cycling together to work.

Miners persisted in leveraging their power as bus consumers. They kept on cycling for months, as long as the weather conditions were good. Bus companies saw their revenues plummet. Some miners even bought a new bicycle and planned on cycling until they recouped the money. Miners who were physically able, kept on cycling to work and boycotted buses. The *Nieuwe Venlosche Courant* reported how whole "caravans of cycling miners can currently be encountered on the roads to mines and homes."[27] Some bus companies, struggling to avoid bankruptcy, were persuaded to accept the old fares, risking the withdrawal of their concession. But there were also reports of miners moving closer or temporarily renting an apartment near their workplace—fed up with the struggle for accessible buses and representation.[28]

Conclusion

"Protesting Bus Regulations" has provided a glimpse of workers' resistance against the dominant mobility regime. These unionized miners politicized the bicycle as both a mobility alternative and a critique against bus companies and the state. The bicycle also symbolized miners' resilience to ensure their mobility access to jobs and their resistance to government-imposed regulations.[29] In other words, the miners showed their agency as workers in shaping their mobility as they militated against the centralizing, technocratic state and bus companies. State regulation of the burgeoning buses (i.e. abundance of cheap mobility options) separated the desired effect of order from the political establishment perspective, but raised cost barriers (mobility poverty) for workers reliant on buses for accessing jobs. As a political statement, the boycotts might have been successful. But like in the South African and Italian case studies, the power to (re)direct transport policies remained with bus companies and the state.

In the following decade, marked by war and scarcity, the precariousness of workers' mobility was even more pronounced. This time, workers had less power to take action. The state and employers intervened.

27 "De vervoerkosten van en naar de mijnen," 8.

28 Ibidem.

29 The miners' cycling protest in 1938 was not part of a bicycle counterculture against the car, campaigns for improved bicycle infrastructure, cyclists' road safety or environmental justice, which are more widely discussed in bicycle historiography. Zack Furness, "Biketivism and Technology: Historical Reflections and Appropriations," *Social Epistemology* 19, no. 4 (2005): 401-402.

3 Mobility Austerity during War and Scarcity, 1940-1947

Amid war and shortages, blue-collar workers experienced severe mobility barriers to access jobs. Europe in the 1940s was marked by shared experiences of war, destruction, and scarcity. Human loss is estimated at around thirty-six and half million Europeans—many of them civilian casualties in bombed out cities and death camps. Widespread shortages of foodstuffs, textiles, and other necessities put a heavy burden on citizens' shoulders. The material damage was enormous too. Across Eastern and Western Europe, cities, homes, and industries lay in ruins, while transport and communications were severely disrupted. In France, of the 12,000 pre-war locomotives, only 2,800 were still running. Roads, rail tracks, and bridges had been destroyed by retreating German forces and advancing Allied troops. In the Netherlands, pre-war rail, road, and canal transport was reduced to 40 percent by 1945.[1] The daily commute was far from self-evident.

Wartime and postwar shortages put workers' already precarious mobility under further pressure. Mobility historiography indicates that road and rail transport was severely affected by wartime damages, confiscations, and shortages, but does not detail how this affected workers' ability to access jobs. "Mobility Austerity during War and Scarcity" argues that workers' agency to overcome mobility barriers was more limited in this period, though where possible, they attempted to negotiate their mobility via employers. As industries were assigned for the war economy, I show that in contrast to the interwar period, workers' mobility became a direct task for the government and employers, who took over the coordination of the everyday commute. This led to a shift in the locus of control over workers' mobility. It also meant that workers in so-called vital industries enjoyed a relative advantage compared to others.

Despite wartime hardships, many aspects of people's daily lives continued during the war, albeit with constraints and sometimes under new, improvised

1 Tony Judt, *Postwar: A History of Europe Since 1945* (London: Random House, 2010 [2005]), 13-14, 16-17.

Bek, Patrick, *No Bicycle, No Bus, No Job: The Making of Workers' Mobility in the Netherlands, 1920-1990*. Taylor & Francis Group 2022
DOI: 10.5117/9789463723183_CH03

conditions. Some even thrived with more disposable income in times of austerity. The daily journey to work was one such aspect of everyday life, though constrained by widespread scarcity and state-imposed mobility austerity. This chapter discusses how resource scarcity and austerity weighed on workers' mobility access to work. The war and occupation were not situations in which dockers, miners, and steelworkers could organize through their unions and leverage their purchasing power and create their own initiatives like before. Still, they had some political leverage as essential workers for the German war industry.

War created a new power dynamic. When the Netherlands capitulated in May 1940, the German occupier established a civil administration. With the Dutch queen and council of ministers in exile in London, Reich Commissioner Arthur Seyss-Inquart fulfilled the role of head of state, leading four German commissioners-general for Public Administration and Justice, Finance and Economy, Public Security, and Special Affairs. These Hitler-approved commissioners directed the Dutch secretaries-general and the ministries' civil servants. Co-operation between the German and the Dutch administration was complicated, vacillating between outright collaboration and subtle resistance. Dutch officials largely remained loyal to the Dutch people and the government in exile, while seeking to appease German authorities as the best strategy to prevent stronger repression. Dutch businesses found themselves in a similar position as they had to comply with German orders when Dutch production became linked to the German war effort. Textile factories were ordered to produce for Germany, steelworkers were forced to produce for the Wehrmacht, and the full mining capacity came under German rule. Dutch managers tried to keep business running as usual for the Dutch public interest—preventing an economic crash, mass unemployment, and scarcity of necessary provisions and resources like coal—while producing for the occupier.[2] Especially after trade unions had been banned in 1942, employers proved important intermediaries for German austerity plans and helping workers overcome mobility barriers.

The predominant response to 1940s resource scarcity, catering to military needs, was austerity for civilians. German authorities put a heavy burden on citizens by appropriating all scarcely available resources for the war machine.

2 Frank van Buren, "Kolenwinning onder Duitse druk: productie- en arbeidsverhoudingen in de Limburgse mijnen tijdens de bezetting 1940-1944," *Studies over de sociaal-economische geschiedenis van Limburg* XLIV (2004): 135-164; Wichtert ten Have, *1940: Verwarring en aanpassing* (Houten/Amsterdam: Spectrum/NIOD, 2015), 84-88, 129-135; Hein Klemann, *De Nederlandse economische contacten met Nazi-Duitsland en de Nederlandse economie tijdens de Tweede Wereldoorlog: Een bibliografie* (Amsterdam: NEHA, 1995).

Ordered by German authorities from the early onset of war, local governments rationed food, textiles, and fuel. Amid this austerity, civilians were resourceful. Historians Ruth Oldenziel and Heike Weber detailed how war and postwar shortages reinforced practices of maintenance, repair, and reuse. These practices continued long after the war. Oldenziel and Milena Veenis characterize the postwar decade as a "war economy in peacetime," marked by scarcity, household thrift, and government austerity.[3] This applied to workers' mobility too.

"Mobility Austerity during War and Scarcity" provides glimpses of how material shortages affected everyday mobility and the government and employers' attempts to manage workers' commute. Perhaps unsurprisingly, lacking their unions which had been suspended by the Germans, workers had little agency as consumers in resolving their mobility poverty in the 1940s. Via their employers they could, however, negotiate their commuter needs. The German occupier operated closely with Dutch industrial employers to feed the German war machine, implementing austerity and rationing policies to distribute the scarcely available mobility resources over society. In this form of governmentality, key questions were: what resources should be saved and redistributed to whom? And who is considered worthy of government support? German austerity policies aimed to control the distribution of scarce resources for the military and war economy, including top-down rationing mobility resources, envisioned to benefit the war economy. How did the government and industrial employers respond to wartime and postwar mobility challenges?

3.1 Wartime Transport Mismatch

The Dutch transport system was under great pressure during the war. In 1940, the German occupying forces prohibited fossil-fueled private motor usage to benefit the military. Coupled with fuel rationing and rubber tire shortages for private citizens, road transport suffered severely. The number of bus trips dropped by 68 percent. As passenger numbers only declined by 15 percent, the few available buses were overcrowded. Mobility historians

3 Ten Have, *1940: Verwarring en aanpassing*, 122-124; Ruth Oldenziel and Milena Veenis, "Het huishouden tussen droom en werkelijkheid: oorlogseconomie in vredestijd, 1945-1963," *Techniek in Nederland in de twintigste eeuw. Deel 4: Huishoudtechnologie, medische techniek*, ed. Johan Schot, et al. (Eindhoven/Zutphen: Foundation for the History of Technology /Walburgpers, 2001): 103-131, here, 103, 105, 107; Ruth Oldenziel and Heike Weber, "Introduction: Reconsidering Recycling," *Contemporary European History* 22, no. 3 (2013): 347-370, here 349, 363-365; Erik Schumacher, *1942: Oorlog op alle fronten* (Houten/Amsterdam: Spectrum/NIOD, 2017), 83-91.

note that public buses were mostly used for short distances. Buses were also vital for worker transport over longer distances. The decline in motorized road transport led to a temporary revival of rail-based mobility in the Netherlands. As tram occupancy soared, companies even re-deployed previously discontinued steam trams to meet the growing demand. By 1940, the number of tram passengers had returned to 1934 levels, and further tripled in the next three years. Eighty percent of tram trips were local. Dutch Railways covered long-distance trips, doubling passenger numbers between 1941 and 1943, after a decade of loss.[4]

When the Second World War broke out, bicycles fulfilled a vital function in everyday mobility for people from all class backgrounds. Across Europe, people used bicycles to get around. Like in Antwerp, Copenhagen, and other European cities, Lyon's citizens turned to cycling when authorities rationed gasoline, as Maxime Huré explains.[5] And in the last year of the war, city residents also used bicycles to travel to the countryside in search of food. In France, that war-time cycling experience would be negatively associated with war and poverty—long after the end of the conflict, as Cathérine Lavenir discovered in her oral histories.[6] In the Netherlands, people were forced to travel greater distances. Housewives who before the war got by in their own neighborhoods, now had to cycle many kilometers to get primary foodstuffs, according to Els Blok.[7] Generally, socialist newspaper *Het Volk*—taken over by Dutch national-socialists during the war—wrote in 1941, "for most of us, the bicycle has become our most precious possession." After German forces invaded the Netherlands in May 1940, the Dutch also rediscovered the bicycle as a "reliable and essential helper through all the hard times," according to a newspaper.[8]

Once again, bicycles proved vital to get around. The bicycle's popularity, however, also aggravated the demand for rubber tires, already under pressure because the military caused global rubber shortages from 1941/42. The Japanese occupation of Malaya and the Dutch East Indies from 1942 drastically

4 Gijs Mom and Ruud Filarski, *Van Transport naar Mobiliteit: De Mobiliteitsexplosie 1895-2005* (Zutphen: Walburgpers, 2008), 240-243.

5 Maxime Huré, "Lyon: The Bicycle Goes Corporate," *Cycling Cities: The European Experience*, ed. Ruth Oldenziel et al. (Eindhoven: Foundation for the History of Technology, 2016): 173-183, here 176-177.

6 Cathérine Lavenir, "Scarcity, Poverty, Exclusion: Negative Associations of Bicycle's Uses and Cultural History in France," in *Cycling and Recycling: Histories of Sustainable Practices*, ed. Ruth Oldenziel and Helmuth Trischler (New York/Oxford: Berghahn Books, 2016), 58-71.

7 Els Blok, *Loonarbeid van vrouwen in Nederland 1945-1955* (Nijmegen: SUN, 1978), 33-34.

8 "Het belang van een goede stalling," *Het Volk*, 1 December 1941, 3.

cut off rubber supplies for the United States and European countries heavily reliant on rubber imports for bicycle and car tires, shoe soles, and other products. The shortages soon put a heavy strain on cycling mobility across Europe. In Belgium, when the German occupier rationed tires, fewer people could cycle; in Copenhagen, residents reverted to walking or public transit, and in the Dutch port of Rotterdam, trams became overcrowded too.[9] The German authorities took note. The German Commissioner for Finance and the Economy (Generalkomissar für Finanz und Wirtschaft), believed in 1941 that the biggest challenge with worker transportation was the "catastrophic tire situation" ("katastrophale Reifenlage").[10]

Workers turned to their employers. In the energy and steel sectors deemed vital for the war economy, workers in particular felt they had leverage to raise the issue. In May 1942, a miner who lived in Eygeshoven and worked at Oranje-Nassau Mine II, complained to the personnel department that he desperately needed new bicycle tires: "Do you realize that I live at least 5 km from the O.N.II mine and so you surely can't expect me to walk all that way, when shoes are so scarce." Again and again, he requested new tires, but it would take another four months before he could get the coupons: "my tires will never last that long," he complained.[11] He was not the only one. A mine inventory showed thousands of miners faced these problems. From a total of 3,343 bicycles counted at all four Oranje Nassau Mine (ONM) locations, about a third of the tires were so worn-out that they had to be replaced within a month. Other miners used "half-worn tires that had to be replaced within six months." Records from 1942 reveal that ONM could only distribute 8.3 percent of the requested outer tubes and 16.7 percent of the inner tubes.[12] Stringent tire rationing presented workers and employers with an impossible challenge.

The bicycle could no longer be a life-saving tool. The workers' already precarious mobility needs faced even greater constraints—unless they worked in industries deemed vital by the war authorities. Chair of the Miners Transportation Committee, engineer M.W.E.E. Reinards, recalled

9 Eric Berkers et al., *Cycling Cities: The Rotterdam Experience* (Eindhoven: Foundation for the History of Technology, 2019), 23; Martin Emanuel, "Copenhagen: Branding the Cycling City," in *Cycling Cities: The European Experience*, ed. Oldenziel et al., 77-87, here 81; Frank Veraart, Stijn Knuts, and Pascal Delheye, "Antwerp: Cycling Claims a Comeback," in *Cycling Cities: The European Experience*, ed. Oldenziel, et al., 89-99, here 93.

10 NIOD, Archive. no. 039, Inv. no. 673, Correspondence 6 July 1942.

11 Regionaal Historisch Centrum Limburg (RHCL herafter), Archive no. 17.05W, Inv. no. 26, Correspondence 28 May 1942.

12 RHCL, Archive no. 17.05W, Inv. no. 26, Correspondence 8 August 1941, Correspondence 27 August 1941, and Internal Minutes 30 October 1942.

in *Limburgsch Dagblad* (1953), that the transport situation during the war and its aftermath was "catastrophic."[13] Bicycle tires were only part of the problem. Fuel and other material shortages also affected workers' mobility. The State Traffic Inspectors wrote in October 1942 that only 25 percent of the necessary fuels could be distributed, severely hampering workers' road transport. The lack of spare parts also drastically reduced the number of operational buses. Florent Habets, owner of Limburg bus company De Valk, petitioned the Secretary-General of the Department of Public Works about this "dire situation." His drivers normally transported 1,600 miners on eighteen buses, but now had to cram 80 miners into buses meant for forty passengers and were only allowed to take miners from 10 km or greater travel distances. "I can hardly imagine what it must be like for these poor men in wintertime, having to struggle day in day out 15, 16, 18, or 20 kilometers through the snow—like so often happened last winter."[14]

In the vital steel industry, low-paid workers also struggled to get to work. In February 1943, steel company Hoogovens received petitions from employees regarding the discontinuation of buses. Seventeen signatories from Uitgeest pleaded for continuation of bus services for several reasons. First, it was well over an hour's walking distance. Second, walking such distances was virtually impossible with "no shoes, because they are always needing repairs." Third, alternatives to walking were lacking. Without the buses, they could not or only with "great difficulty" get to the blast furnace sites and work round-the-clock shifts.[15] A worker from Heemskerk pleaded for buses because otherwise he had to walk 17 km, that is 3 hours back and forth. He "couldn't buy tires, and shoes were becoming scarce … and there is no other means of transport."[16] Many steelworkers complained that without buses, they saw no chance of getting to work on time, emphasizing this was not in their nor the company's interest.[17]

For workers and their families, not being able to commute meant they could not make a living. For industries it meant higher absenteeism rates and

13 "Busvervoer van mijnwerkers tot op de minuut geregeld," *Limburgsch Dagblad*, 16 December 1953, 5.

14 NIOD, Archive no. 039, Generalkommissariat für Finanz und Wirtschaft, Inventory no. 673, Stukken betreffende het vervoer van mijnwerkers van en naar de mijnen, 1940-1942, Correspondence 30 July 1942, Correspondence 5 August 1942, Correspondence 31 October 1942.

15 Tata Steel Central Archives (TSCA hereafter), Archive no. 4350, Vervoer van en naar Heemskerk, Uitgeest, Zaanstreek, Amterdam, Inventory no. RA-12878, Correspondence n.d. February 1943.

16 TSCA, Archive no. 4350, Inv. no. RA-12878, Correspondence 28 February 1943.

17 TSCA, Archive no. 4350, Inv. no. RA-12878, Correspondence 1 March 1943, 2 March 1943, and 5 March 1943.

production loss. War archives include correspondence between bus companies and the German Commissioner, showing that with fewer buses available, workers came in late more often. For workers this meant a loss of wages. The Commissioner warned in 1942 about "a slackening of performance and job satisfaction" and "arbeidsfreude." The Minister of Public Works had similar concerns. He acknowledged in 1942, that "Standing for an extra hour and half every day in crammed-full buses and trams must cause exhaustion, and inevitably lead to high sickness levels and lower productivity."[18] The paternalistically oriented Oranje-Nassau Mines management also worried about the impact on their employees' health and productivity. In a joint effort, mining managers in Limburg wrote to the State Bureau for Rubber in July 1941, that many miners living outside traditional mining towns relied heavily on their bicycles. Soon they would be forced to walk without buses or trains. This physical effort came on top of already physically demanding work. A miner's working day was about 9.5 hours, including preparations before descending into the mine shaft, transport to and from underground workstations, and mandatory baths after shifts. Longer travel times in the late evening or early morning (depending on shifts) on foot would further reduce "the rest that miners need after a heavy day's work" and would "have a negative effect on the peace and quiet they need to recover from their labor."[19] The mobility situation also negatively interacted with other facets of workers' daily lives. In co-operation with Dutch employers, German authorities sought solutions to keep the war industry going.

3.2 Wartime Mobility Austerity

People in all walks of life personally experienced the impact of war-time shortages. Only some received state support for their mobility needs. In collaboration with industrial employers, the wartime government implemented a centralized social triage to promote economic stability and industrial production. Workers who did not work in what the nation considered crucial sectors had to fend for themselves. Workers in vital industries were to some extent supported in their commute. Not by unions like in 1938, but by the state and employers. Across occupied Europe, (military) authorities

18 Nationaal Archief (NA hereafter), Archive no. 2.16.84, Ministerie van Waterstaat: Afdeling Vervoerwezen, (1938-)1941-1945(-1952), Inv. no. 601, Vervoer van arbeiders, 1942, 18 December 1942, 2; NIOD, Archive. no. 039, Inv. no. 673, Correspondence 30 July 1942, 5 August 1942.

19 RHCL, Archive no. 17.05W, Oranje-Nassau Mijnen: Oorlogstijd en na-oorlogstijd, 1936-1966, Inv no. 26, Distributie van rijwielen, rijwielonderdelen en rijwielbanden voor personeel Limburgse Mijnen, Correspondence 1 July 1941.

privileged certain social groups in distributing scarce food, clothing, and other items. German authorities considered some industrial workers essential for the wartime economic effort. Heavy industries like coal and steel depended on a mobile labor force deemed essential for their proper functioning, which German authorities marked as "extremely important for the war effort" ("außerordentlich kriegswichtig").[20] Workers in these industries were also exempt from German bicycle ordinances and prioritized in the allocation of scarcely available resources like bicycle tires.

These bicycle ordinances were significant in the Dutch collective memory of the war years. The impact on workers' mobility seemed modest. The German Commissioner for Finance decreed in 1942 that exempt from the ordinance were "all workers working in companies, which of course also includes employees and those who work in agriculture."[21] Workers in vital industries like the mines, docks, and steelworks received bicycle dispensation from the German occupying authorities.[22] Frank Veraart shows that in mining town Kerkrade, 2,095 miners, 1,067 workers in other (smaller) industries and trade, 396 service workers, and 147 agricultural workers received a bicycle dispensation.[23] Citizens could not simply escape resource scarcity. It required another government strategy: mobility austerity.[24]

In the 1940s, government austerity measures entailed (re)allocating scarce resources to specific groups of people. Determining who should be allocated resources, was a process of categorization and social differentiation, in which the government and industrialists set the norms for who had priority.

20 NIOD, Archive no. 039, Generalkommissariat für Finanz und Wirtschaft, Inventory. no. 1064, Correspondence 13 October 1943, Correspondence 20 August 1943; Judt, *Postwar*, 21.

21 NIOD, Archive no. 039, Inv. no. 1064, Stukken betreffende de invordering en toewijzing van auto's, benzine, fietsen en banden aan bedrijven, Correspondence 20 July 1942.

22 NIOD, Archive no. 249-0881A, Dossier Vrijstellingen, Inventory no. 1, Bundel vrijstellingsbewijzen voor de rijwielvordering en verklaringen van werkgevers enz. betreffende de onmisbaarheid van een rijwiel, afgegeven in verband met de rijwielvordering, 1940-1945, Correspondence n.d. Augustus 1942 (Rotterdam), Vrijstelling vordering van rijwielen, n.d. ("N.V. Philips' Gloeilampenfabrieken"), Vrijstelling Rijwielvordering, 10 September 1942 (Mines, Heerlen), Vrijstelling, 29 Januari 1945 (Hoogovens), Correspondence 20 September 1944 (Hoogovens), Correspondence 28 Juli 1942 (Hoogovens); RHCL, Archive no. 17.05H, Oranje-Nassau Mijnen: vervoer, 1903-1978, Inv no. 50, Vordering rijwielen in oorlogstijd, Correspondence 19 juli 1942, Correspondence 20 Juli 1942; see also Jordan's observations regarding bicycle confiscations in his popular historical work on cycling in Amsterdam: Pete Jordan, *De Fietsrepubliek* (Amsterdam: Uitgeverij Podium, 2013).

23 Frank Veraart, "Cycling in Numbers: 1920-2015," in *Cycling Cities: The European Experience*, ed. Oldenziel et al., 200-229, here, 215.

24 For discussion of the concept of "mobility austerity" see Anna Nikolaeva et al., "Commoning Mobility: Towards a New Politics of Mobility Transitions," *Transactions* 44, no. 2 (2019): 346-360.

Employers mediated this process: identifying workers' mobility barriers, communicating their needs to state officials, and seeking solutions tailored to workers' precarious situation. At the same time, employers tried to gain greater control of workers' movements between home and work, reducing any liabilities for production through mobility austerity.

Paternalistic employers put their full weight behind ensuring their workers could get to work through a range of emergency and permanent measures. Dutch companies first lobbied the German Commissioner for allocation of material resources. As an emergency solution, employers often paid the difference in travel costs when fares were raised for their employees. They continued to lobby for adjusting timetables to working shifts. Some large employers—like electronics maker Philips and the Oranje-Nassau Mines continued interwar practices—paying employees' travel costs and modest bicycle allowances. New during wartime was that the government provided special "route tickets" for workers in "important war operations" ("ein kriegsgewichtiger Betrieb").[25] Employers mediated between workers and the state's distribution authorities. In Limburg, a special overarching body was established, the Miners Transportation Committee, consisting of state inspectors, bus companies, mining companies, and union representatives. In most other cases, employers cooperated closely with the state in handling mobility austerity and distribution for civilians. The authorities' austerity policy involved using available mobility resources for civilians more efficiently, allocating these to designated industries and workers, and reducing any mobility-related liabilities for industrial productivity.[26]

25 NIOD, Archive. No. 039, Inv. No. 1102, Stukken betreffende de mogelijkheden tot het gebruik van tram, trein of bus door arbeiders, 1942-1944, Correspondence 25 October 1943.

26 NIOD, Archive no. 039, Inv. No. 1065, Stukken betreffende de invordering en toewijzing van auto's, benzine, fietsen en banden aan bedrijven, Correspondence 13 April 1944 and 5 April 1944, Correspondence 27 September 1943; Philips Company Archives (PCA hereafter), Inv. No. 631.4, No. 1, Arbeidsvoorwaarden 1 January 1944; RHCL, Archive no. 17.05H, Inv. no. 18.1, Correspondence 2 July 1942, Announcement 8 October 1942; RHCL, Archive no. 17.05H, Inv. no. 46, Announcement 7 June 1941, Correspondence 6 September 1941; TSCA, Archive no. 4337, Inv. no. RA-04155, Minutes 22 October 1942, 2; "Rijwielvergoedingen," *Philips Koerier*, 24 February 1943, 3. Both the government and employers compensated for workers' high travel costs. In 1941 mining companies paid the difference between old and new fares. Hoogovens Social Affairs sold special bus tickets for ten trips along designated routes and reduced fares between 1.05 and 1.95 guilders. In 1943 correspondence with companies, the German Commissioner for Economy and Industry decreed that workers' travel costs were to be partly compensated by the state. Workers could apply for a special route tickets ("Trajektkarte") when working in a critical industry. Personnel magazine *Philips Koerier* read in 1943, that Philips employees received a bicycle allowance—ranging between an annual 10 and 50 guilders—from the Salary Department in case workers could not perform their work tasks without bicycle or lived at more than three-kilometer distances from work.

Workers in larger industries were represented by their employers in the centralized distribution of bicycle and tires via regional distribution offices. They filled in a form—name, place of residence, distribution card number, and details of the materials needed—and answered specific questions about their job, position, and nature of the company's operations. Workers in companies with more than ten employees had to file requests via their managers. A collective request on behalf of the company was then forwarded to the distribution authorities.[27] Workers in either companies with fewer than ten employees or not designated as critical industries—and people not employed at all—had to fend for themselves. The authorities tasked to keep the economy running for the war industry could not honor most requests—not even from industries producing for the German war effort. It forced wartime authorities to come up with new categories to determine who should receive support for their mobility needs or who they considered was creating value.

Companies also created different categories based on how far workers traveled—taking the location of the factory or mining site as point of departure—and a worker's position. In May 1941, miners at the Oranje Nassau Mines could request bicycle tires from their manager, but only if they could prove their old tires were worn and, most importantly, they needed their bicycle to commute to the mines or a bus or train stop near their home. This travel distance had to be at least 3 km. But since most miners lived within this radius, they were low priority and expected to fend for themselves. At the Oranje Nassau Mines, additional positions "in the interest of the company" increased the likelihood of qualifying for bicycle tire coupons, for example members of the fire or rescue brigade, gas scouts, or mine police.[28]

Other companies set similar requirements with high thresholds to qualify for bicycle tire coupons. In October 1942, Hoogovens employees received a letter "Distribution according to place of residence," notifying them that, in order to efficiently organize transport to and from the blast furnaces with the scarce resources in the winter of 1942-43, they had to categorize workers as: Category A (about 200) lived in places below sufficient accessibility and would use company-chartered autobuses. Category B workers did not need to be included as they lived in places like Castricum (12 km from Hoogovens) and were expected to travel to nearby Uitgeest (3.5 km east) before using

27 NIOD, Archive. no. 249-0198G, Dossier – Distributie – Rijwielen, Inv. no. A1, Bundel modellen van formulieren, de distributie van rijwielen en banden betreffende, 27 Juli 1940-25 September 1946; "Distributie: Fietsbanden en fietsen," *De Tijd*, 14 February 1947, 2.

28 RHCL, Archive no. 17.05W, Inv. no. 26, Correspondence 1 February 1941, Announcement 16 May 1941, Correspondence 16 May 1941 and Appendix.

public transit services. Category C workers (about 125) were granted access to company buses. So-called D workers had to travel by train on their own initiative, and category E workers in IJmuiden and Velsen were expected to use the ferry (when ice drift did not prevent the ferries from running).[29]

A meeting with Philips Works Department and Distribution Center shows that only employees below the sufficient accessibility threshold received support—that was set at five km instead of three, and only if there were no mobility alternatives, according to the Nazi Arms Inspection (Rüstungsinspektion Niederlande), personnel magazine *Philips Koerier* (1943) reported. Philips workers were also prioritized for other reasons, namely: employees who used their bicycle "very frequently for greater distances" during shifts, workers with a doctor's statement that they could not walk to the factory, employees with ancillary positions in security, "Air Security Service," fire rescue, and medical services. Such employees received a written "employer's declaration," stating they qualified for bicycle tires at the local distribution office.[30] These austerity measures continued after the war, as widespread shortages still hampered workers' commute.

3.3 Scarcity and Austerity Continue After the War

In *Postwar*, historian Tony Judt defined the era between 1940 and 1947 as "the Age of Austerity": shortages and austerity measures did not end with the war. Intense destruction in the final year of the war left both sides in ruins. He characterized the final war year as "utter misery and desolation" with "pitiful streams of helpless civilians trekking through a blasted landscape of broken cities and barren fields." Not only people but also things like transport technologies seemed "worn out, without resources, exhausted," he wrote. "Even trams, propelled uncertainly along damaged tracks by intermittently available electric current, appear shell shocked."[31] Almost everything from foodstuffs to clothing and fuel was either rationed or unavailable. In France, the worst scarcities were resolved around 1949, in Britain, food rationing only ended in 1954. In the Netherlands, the last rationed product (coffee) became freely available by 1952.[32]

29 TSCA, Archive no. 4337, Inv. no. RA-04155, Minutes 22 October 1942, 1-3.

30 PCA, Archive no. 642.11, Inv. no. 1, Internal Minutes 14 May 1943, 2-3, Correspondence 30 September 1943; "Rijwielbanden!," *Philips Koerier*, no. 1 (1943): 1.

31 Judt, *Postwar*, 13.

32 Judt, *Postwar*, 162, 253.

The liberation began disappointingly in the Netherlands too: much of country was left damaged, empty, and destitute with a half-starved population in its major cities. Most people working in Dutch industries were exhausted after the war, but still struggling with daily hardships and shortage of foodstuffs and clothing. Company physicians in Twente's textile industries found workers underweight, anemic, and "anxious or suffering from exhaustion after work," as historian Nick Vos cites contemporaries in his research on postwar labor relations.[33]

Material shortages continued to constrain workers' mobility after the war. The Catholic Miner's Union wrote to the president-director of the Dutch Coalmines (Nederlandse Steenkolenmijnen) in 1947 that the "poor travel facilities" had reduced the incentive for mining because the job caused "fatigue, lethargy and sickness," causing "absenteeism, reduced production, and stagnation on various work floors."[34] And in the words of Gijs Mom and Ruud Filarski, the war had "literally and figuratively carved deep crevasses in the Dutch road network." In the final war months, retreating German forces had flooded large tracts of land, destroyed more than 900 bridges for road traffic and 180 railway bridges, and looted railway materials including hundreds of kilometers of rails. Allied bombers had tried to cut off retreating German soldiers by destroying road infrastructures. The material damage was greater in the Netherlands than other European countries. Postwar recovery was slow due to material shortages.[35]

The postwar damage also impacted workers' mobility. According to former traffic inspector-general, Ir. Th. M. B. van Marle in *Domestic Transportation After the Liberation* (*Het binnenlands vervoer na de bevrijding,* 1953), the postwar mobility system was in crisis: on the one hand there was a serious "lack of means of transport" and on the other, a "great need for transport." The railroad and tram expert particularly observed how workers "who used to travel by bicycle no longer possessed this means of transport." They needed the bus, while "for greater distances, the Railways failed too."[36] Miner Transportation Committee minutes give a sense of the problems

33 Nick Vos, *De rauwe wet van vraag en aanbod: Arbeidsverhoudingen in de Twents-Gelderse textielindustrie, 1945 tot 1949* (PhD Dissertation, University of Amsterdam, 2011), 204, 369-370.

34 RHCL, Archive no. EAN0065, Algemeene Bond van Christelijke Mijnwerkers in Nederland te Heerlen (1908-1940) en de Nederlandse Katholieke Mijnwerkersbond (NKMB) en voorgangers te Heerlen (1907-1975), Inv. no. 1297, Stukken betreffende klachten, 1944-1956, Correspondence 30 September 1947.

35 Mom and Filarski, *Van Transport naar Mobiliteit* (*II*), 245.

36 Th. M. B. van Marle, *Het binnenlands vervoer na de bevrijding: 1944/45-1952* (The Hague: Nederlands Verkeersinstituut, 1953), 44, 49-51.

facing miners. In 1946, union chairman Mulder claimed that due to "late or no buses at all, miners are losing half an hour every day," resulting in serious wage losses. Miners from Stevensweert could not get to work for two days, and miners in Vincken, Van Ool, Richter, Welp, Rutten, and Ramakers waited for buses in vain. Mulder thought this was particularly problematic, because "the majority were married men," wage earners who had to provide for their families.[37] That these men were breadwinners was used as argument to prioritize their mobility.

Improvised "emergency buses" were a welcome alternative. The Ministry of Public Works' Bus Agency (Bureau Autobussen) handled requests for group transport that fell outside the public transit domain. And it disapproved of the trucks and emergency buses for transporting miners, factory workers, agricultural workers, and unemployed men working on government work projects in the Netherlands. In the interwar years, this state control prevented unbridled proliferation of the bus services competing with railways, but now the agency had to ensure that the scarcely available bus materials were allocated to the industries and workers with the most urgent need.[38] In the textile industry area, Twente's newspaper *Tubantia* wrote in 1948 "people were pleased to bits if they could lay their hands on a truck; this was then converted into suitable passenger transport. Not an ideal solution, but you had to make do with what you had."[39] Employers like the State Mines (fig. 5) and Philips also used tinkered trucks to get workers to and from work.[40] Although emergency buses brought some solace, capacity was still too limited to transport all the workers the employers needed. Moreover, these buses were often laid on by companies, and thus dependent on whether employers were willing or able to pay the costs. According to Vos, when cotton mill Arntzenius Jannink & Co. planned to discontinue the buses

37 RHCL, Archive no. 17.05H, Oranje-Nassau Mijnen: Vervoer, 1903-1978, Inv no. H6, Commissie Mijnwerkersvervoer van het Beheer Nederlandse Mijnen, Internal Minutes 13 February 1946, Internal Minutes 27 February 1946, Internal Minutes 4 September 1946; RHCL, Archive no. EAN0065, Algemeene Bond van Christelijke Mijnwerkers in Nederland te Heerlen (1908-1940) en de Nederlandse Katholieke Mijnwerkersbond (NKMB) en voorgangers te Heerlen (1907-1975), Inv. no. 1297, Stukken betreffende klachten, 1944-1956, Correspondence 22 March 1946, 25 March 1946.

38 NA, Archive no. 2.16.83, Ministerie van Waterstaat: Bureau Autobussen, 1943-1950/ Commissie Vergunningen Personenvervoer, 1937-1959, Inventory no. 61, Stukken betreffende het regelen van arbeidersvervoer door de Rijksdienst voor de uitvoering van werken (1945-1949), and Inventory no. 78, Stukken betreffende gebruik van voor personenvervoer goedgekeurde vrachtauto's en noodautobussen (1945-1950).

39 "Arbeiders wilden niet in een "Rijksbus" rijden: Strijd om arbeidersvervoer in Borne," *Twentsch Dagblad Tubantia*, 17 December 1948, 2.

40 PCA, Archive No. 642.5, Inv. no. 2, Correspondence 21 March 1946.

Figure 5: A Bedford truck repurposed into an emergency bus for transporting miners. N.p., 1945 (Source: Regionaal Historisch Centrum Limburg, Fotocollectie DSM, PR27, Creative Commons: BY-NC-SA)

running to the village of Enter in November 1946, the mill's washer women revolted and put down work for four hours. The washer women lived eight km from the mill, and the company-chartered bus was the only mobility alternative besides a long walk. Discontinuing this bus would raise barriers for getting to work and making ends meet at this tumultuous time. Their strike was effective: Arntzenius Jannink continued to pay for a bus service until further notice.[41]

For employers, workers unable to get to work (in time) equated with counterproductivity. Like during the war, workers in vital industries received growing support from the state and industries to overcome their mobility poverty. With the war just over, management at vital steel company Hoogovens wrote to the traffic inspectors in Amsterdam that "after overcoming so much hardship, we are now ready to run our company again. Yet one major hurdle we still have to solve … the transport for getting workers to and from the factory."[42]

41 Vos, *De rauwe wet*, 204, 369-370.

42 TSCA, Archive no. 4344, Personeelsvervoer per autobus (HFD), Inv. no. RA-05831, Correspondence 2 October 1945.

Figure 6: Hoogovens' bicycle repair men. IJmuiden, 1947 (Source: Tata Steel Central Archives)

And bicycle (tire) shortages were not only a war-time issue. This remained a persistent problem until the late 1940s. Workers accustomed to repairing their own bicycles could not get the necessary spare parts. Until 1947, bicycle repairmen on companies like Hoogovens payroll, tinkered with whatever materials they could get their hands on (fig. 6). The bicycle tire situation was still a big concern. *Algemeen Dagblad* reported in October 1946 that the central distribution offices were behind with the demand for two million bicycle tires, noting this was "nowhere near the number needed" estimated at around three million, excluding "the wear and tear on the tires already distributed."[43] In 1947, tourist organization ANWB, reflecting on "our worrying state of cycling" in their magazine *De Kampioen*: the bicycle tire shortage affected all walks of life in the Netherlands.[44] Philips personnel

43 "Achterstand bij distributie van banden," *Algemeen Dagblad*, 16 October 1946, 1.
44 "Onze zorgelijke rijwielpositie," *De Kampioen* 62, no. 3 (1947): 77-78.

magazine *Philips Koerier* assessed the "bicycle tire problem" thus: out of a total of 20,000 men and women living in and around Eindhoven, devoid of other mobility options, so far only 16 percent had received new bicycle tires.[45]

Stringent rationing of bicycle tires (and other items) continued after the war across Europe. Mobility scarcity was still widespread in people from different social backgrounds, but most profoundly among those already struggling to make ends meet. Philips' higher staff and engineers were urged by management to drive economically with their (company) car in the late 1940s. Workers had to make do with ramshackle bicycles or underserviced buses and trains.[46]

Company records give a sense of the conditions surrounding bicycle tire shortages. Hoogovens employee and communist trade union (Eenheids Vakcentrale) representative P.M. Mulder described the steel workers' bicycles in 1947 as in a "deplorable state" after five years of "riding on massive tires."[47] The Miners Transportation Committee representing miners' interests, had similar concerns. The managers of Willem Sophia Mine in Spekholzerheide needed bicycle tires for about 1,000 workers, who had no mobility alternatives in hilly, isolated villages like Elsloo and Stein (25 km from the mine).[48] Lightbulb factory Philips received numerous requests from workers and their team leaders for bicycles and rubber tires. Worker H.M.L. from Geldrop, who lived about 8 km from the Eindhoven sites, requested two inner and outer tubes to replace his wooden tires. Another worker, who "due to abnormal feet," was unable to walk long distances, requested replacement tires too: "Up till now I have used my own tires, but these are now worn and irreparable."[49] As soon as tires became available, employers honored these requests.

In the meantime, postwar distribution offices applied a similar social triage as during the war. Workers who had what was considered sufficient access—living at walking distance from work or a public transit hub—were largely left to their own devices. Those below that sufficiency level received company support within strictly demarcated boundaries. In February 1947, Catholic newspaper *De Tijd* informed readers about the qualifications for

45 "Rijwielbandenverstrekking," *De Vrije Philips Koerier* 2, no. 40 (1946): 4.

46 PCA, Archive no. 728.14, Afdeling Vervoer en Garage, Inv. no. 3, Internal Communication "Aan alle berijders van een auto!" (n.d.).

47 TSCA, Archive no. 4342, Personeelsvervoer per fiets, bromfiets en motor (HFD), Inv. no. RA-12873, Correspondence 6 May 1947.

48 RHCL, Archive no. 17.05H, Inv. no. H6, Internal Minutes 17 March 1946, Internal meeting 13 February 1946.

49 PCA, Archive no. 642.11, Fietsen en fietsbanden, Inv. no. 1, Correspondence 15 May 1946, Correspondence 15 April 1947.

replacement tires from the central distribution offices: employed and residing at least 3 km from the workplace. However, if workers could commute by public transit or had received tires less than ten months ago, they were ineligible.[50]

Some employers still mediated between workers, government, and distribution offices, though gradually found ways to circumvent the official routes by placing bicycle tire orders themselves abroad. In 1946, Philips for instance successfully placed an order with bicycle tire exporters in Brazil. This did not, however, put a stop to social stratification in distribution. The personnel department started by asking heads of departments to list all employees who qualified for new tires. Then they had to give each request an "urgency rating": 1 marked "needs to be helped as soon as possible," 2 "can only qualify after the urgency group is complete." Philips also drafted lists of Eindhoven streets considered too far away to walk and lacking public transit.[51] Other company managers seemed less inclined to develop such initiatives after the war. At Hoogovens, employee requests for bicycle allowances to cover replacement parts and repairs, were declined. By 1947, management expected all workers to spend 60 cents a week on bicycle repair and maintenance. For workers whose bicycle was in a deplorable condition, Hoogovens provided lease-purchase contracts or deposits to enable workers to cycle to work.[52] In short, employers had become important actors in supporting workers' mobility.

Conclusion

In the Age of Austerity, widespread resource scarcities profoundly reduced civilians' ability to get to work. Workers' mobility access was already precarious in the interwar years. The war and immediate postwar years only exacerbated the situation. The 1940s mobility scarcity and rubber tire shortages highlighted workers' high dependency on bicycles, also revealing the entanglements of Dutch everyday mobility and the global resource circulation of rubber. For the first time the state stepped in. It approached mobility scarcity as a distribution challenge: who was eligible for the scarce

50 "Distributie: Fietsbanden en fietsen," 2.

51 PCA, Archive no. 642.11, Inv. no. 1, Correspondence n.d. 1945, Correspondence 24 January 1946, and Correspondence 7 August 1946

52 TSCA, Archive no. 4342, Inv. no. RA-12873, Correspondence 1946, Request Form Bicycle (April 1947).

materials? The economic status of a company and individual workers became inscribed in wartime austerity measures, rendering 'unproductive' unemployed people invisible or unimportant. Everyone's position in the labor market/economy determined their mobility in this period—and remained so in the following decades.

Workers in vital sectors like agriculture, mining, ship-building, steelworks, and textile industries had some room in negotiating their mobility needs, even though unions had been banned by the German authorities. Employers took up the task as mediators in the practical implementation of state-directed mobility austerity, gradually shifting towards central governance. This was a trend in other domains of public governance as well. In the postwar corporatist model, major employers maintained a strong role in national policymaking. They also had a vested interest. Postwar reconstruction and the push for rapid industrialization meant industries increased production. Reconstruction also meant there was plenty of work in Dutch factories and mines, but how were people supposed to get to the workplace? The next chapters detail the mobility challenges that workers and employers faced in the postwar decades. They show how employers soon became a driving force in identifying and resolving workers' mobility barriers.

4 Mobility Barriers during Postwar Industrialization, 1947-1970

The postwar push for heavy industry created high demand for a mobile labor force. Philips worker transport manager P. Dekker recalled in 1953, that besides the shortage of materials and machines, there was a huge shortage of unskilled and semi-skilled workers, and mobility challenges had to be solved.[1] Like in previous decades, the postwar decades' blue-collar workers had to overcome a spatial mismatch between their homes and job locations. Labor historians mention that employers were confronted with these challenges, but do not detail the specific barriers.[2] "Mobility Barriers during Postwar Industrialization" argues that workers' (im)mobility was addressed as an issue of mobilizing workers for vital industries between 1947 and 1970—not for a war economy, but for postwar reconstruction and industrialization. The Dutch government built transport infrastructures and subsidized public (rail) transit as mobility historians have shown, but did not develop a vision on workers' mobility like employers. Mobility scholars attribute an important role to (local) governments in identifying and reducing mobility barriers for low-income citizens.[3] Even though rapid industrialization was a national policy, it was not the government, but paternalistically oriented employers who identified workers' mobility barriers in order to make the everyday commute governable.

1 Philips Company Archives (PCA hereafter), Archive no. 642.5, Inv. no. 882, P. Dekker, "Rapport aan de Minister van Verkeer en Waterstaat inzake de N.V. tot Vervoer van Industrieel Personeel Rayon Eindhoven (V.I.P.R.E) te Eindhoven" (1953), 2.

2 Bram Bouwens et al., *Door staal gedreven: Van Hoogovens tot Tata Steel, 1918-2018* (Bussum: Uitgeverij Thoth, 2018), 119-120; Serge Langeweg, *Mijnbouw en arbeidsmarkt in Nederlands-Limburg: Herkomst, werving, mobiliteit en binding van mijnwerkers tussen 1900 en 1965* (Hilversum: Verloren, 2011), 63-72; Erik Nijhof, *'Gezien de dreigende onrust in de haven': De ontwikkeling van de arbeidsverhoudingen in de Rotterdamse haven 1945-1965* (Amsterdam: IISG, 1988), 138-142; Jan Zwemer, "De pendelarbeiders van Tholen en Sint-Philipsland," in *Zeeland 1950-1965*, ed. Jan Zwemer (Vlissingen: Uitgever Den Boer/De Ruiter, 2005), 401-428.

3 Tobias Kuttler and Massimo Moraglio, *Re-Thinking Mobility Poverty: Understanding Users' Geographies, Backgrounds and Aptitudes* (London/New York: Routledge, 2020), 14.

Bek, Patrick, *No Bicycle, No Bus, No Job: The Making of Workers' Mobility in the Netherlands, 1920-1990*. Taylor & Francis Group 2022
DOI: 10.5117/9789463723183_CH04

Like in surrounding countries, Dutch postwar reconstruction centered around rapid industrialization, formalized in the first Industrialization Bill (*Nota inzake industrialisatie in Nederland*) in 1949. The national government kept labor costs low by imposing a guided wage policy and facilitated transportation of people and goods by making infrastructure improvements, funded in part by the Marshall Plan.[4] At this time of economic expansion, companies were eager to locate cheap laborers. According to employers facing labor shortage, young, unmarried women were more suitable for precision work in manufacturing and textile industries, and, more importantly, cost a third less than men. Employers found that this shortage of workers made many Dutch women opt for less dirty and physically demanding work on assembly-lines and in textile laundries, according to Els Blok in *Loonarbeid van vrouwen 1945-1955*.[5] For its low-paid and unskilled work, electronics manufacturer Philips in Eindhoven looked across the border, in Belgium's Campine region, where plenty of young rural women were eager to work at Philips, transport manager Dekker wrote in 1948.[6] But how to get these women to Eindhoven?

Attracting rural and cross-border workers was a general trend. According to historians Erik Nijhof, Melchior van Elteren, Bram Bouwens, and Serge Langeweg, port industries in Rotterdam, Hoogovens in IJmuiden, and the mines in Southeast Limburg attracted growing numbers of workers from distant rural regions in the first post-war decade. The agricultural sector was undergoing rapid mechanization, causing high unemployment among young farm hands with few job prospects. Further afield, migrant workers were attracted by Dutch (and other European) industries to resolve labor shortages. In 1949, the Dutch government signed an agreement to recruit Italian workers for Limburg's mines. In the following decade, textile industries, shipyards, and steel giant Hoogovens persuaded the Ministry of Social Affairs (Sociale Zaken) to run additional recruitment campaigns. Around 1960, like elsewhere in Western Europe, Dutch companies hired migrant workers from Greece, Morocco, Spain, Turkey, and Yugoslavia, who were supposed to come and work on a temporary basis with a special work

4 Greet De Block, David de Kool, and Bruno de Meulder, "Paradise Regained?", in *Builders and Planners: A History of Land-use and Infrastructure Planning in the Netherlands*, edited by Jos Arts et al. (The Hague: Eburon Academic Publishers, 2016), 17-68, here 60.

5 Els Blok, *Loonarbeid van vrouwen in Nederland 1945-1955* (Nijmegen: SUN, 1978), 60-62, 67, 74.

6 PCA, Archive no. 642.5, Inv. no. 882, Correspondence 31 December 1948, 1, 3, and newspaper clipping "Dagelijks volksverhuizingen van België naar hier: Lange rijen autobussen voeren fabriekspersoneel aan," *Het Parool*, 26 October 1949.

permit—performing low-value and unskilled work for low wages compared to Dutch workers.[7]

Another issue was how to get these workers to the docks, factories, and mines. Mobility scholars have indicated that the lack of affordable housing near jobs and failing public transit to get there are major barriers for low-income women and men to land entry-level jobs.[8] Workers and managers faced two main challenges. First, widespread housing shortages meant there was no accommodation for workers near factories and mines. Second, managers claimed public transit failed to meet industry and workers' mobility needs, causing absenteeism and production loss. The chapter shows how the lack of housing and public transit raised barriers for people getting to work.

4.1 Lack of Affordable Housing Near Jobs

Housing these huge numbers of workers near factory and mining sites proved impossible. Many houses had not survived the war. Of the 2.1 million houses in the Netherlands in 1940, almost 87,000 homes were destroyed, 43,500 heavily damaged, and 293,000 lightly damaged. Although the state's postwar policy centered on rapid reconstruction, it could not counter the wartime stagnation of housing construction in the short term. The scale was too large, and the government did not prioritize public housing right away. Postwar reconstruction initially focused on economic recovery, industrial expansion, and improving exports.[9] National governments across Europe faced the same challenge of

7 Bouwens et al., *Door staal gedreven*, 119-120; Melchior van Elteren, *Staal en Arbeid: Een sociaal-historische studie naar industriële accomodatieprocessen onder arbeiders bij het desbetreffend bedrijfsbeleid bij Hoogovens IJmuiden, 1924-1966* (Leiden: E.J. Brill, 1986), 783, 887; Langeweg, *Mijnbouw en arbeidsmarkt*, 63-72; Nijhof, *Gezien de dreigende onrust*, 138-143; Jan Peet, "Werken en leren 1948-1973," in *Vertrouwde patronen, nieuwe dromen: Nederland naar een modern industriële samenleving 1948-1973*, ed. Hélène Vossen, Marjan Schwegman, and Peter Wester (IJsselstein: VGN, 1992): 122-127, here, 123; Jaap Vogel, *Nabije vreemden: Een eeuw wonen en samenleven*, Cultuur en Migratie in Nederland (The Hague: SDU Uitgevers, 2005), 83-84, 140.

8 Jeroen Bastiaanssen, Daniel Johnson, and Karen Lucas, "Does Transport Help People to Gain Employment? A Systemic Review and Meta-Analysis of the Emperical Evidence," *Transport Reviews* 40, no. 5 (2020): 607-628; Kuttler and Moraglio, *Re-Thinking Mobility Poverty*; Kate Morris, "Research into travel horizons and its subsequent influence on accessibility planning and demand responsive transport strategies in Greater Manchester," paper presented at the *European Transport Conference*, Strasbourg, France (2006); Dominic Villeneuve and Vincent Kaufmann, "Exploring the Causes of Social Exclusion Related to Mobility for Non-Motorized Households," *Transportation Research Board* 2674, no. 8 (2020): 911-920.

9 Leon Deben and Ineke Teijmant, "Wonen is meer dan een dak boven je hoofd," in *Vertrouwde patronen, nieuwe dromen*, ed. Vossen, Schwegman, and Wester (IJsselstein: VGN, 1992), 134-141,

resolving housing shortages, according to historians Kees Schuyt and Ed Taverne. In the Netherlands this was no different. Grand plans had been sketched already during the war, but these proved difficult to implement. While a welfare state was created, public housing fell by the wayside. Like in surrounding countries such as Denmark, Britain, France, and West-Germany, the Dutch government envisioned housing construction could be left to the market once material damages were repaired and construction companies up and running. In 1962, the Dutch still considered the housing shortage "public enemy number one."[10] Affordable housing near (entry-level) job locations—a precondition to preventing mobility poverty problems, scholars have argued—was a postwar issue in the Netherlands. The housing and commuter problem was not new. It did exacerbate. Before the Second World War, employers had recruited migrant workers, and solved the housing and commuting issue by offering company housing nearby. The war-time destruction of housing prevented Dutch industrial employers from doing this on a large scale in the postwar years.[11]

State-led campaigns to attract migrant workers put even more pressure on the already problematic housing situation. Between 1949 and 1975, tens of thousands of Italians found work in Dutch industries, followed by Spanish workers, who were escaping rural poverty and fleeing the Franco regime. Between 1960 and 1973, tens of thousands of migrants from Central and Southeast Turkey arrived in the Netherlands and thousands of Moroccan workers—around 4,000 via official government recruitment channels, and many more by their own means. By 1973, when official recruitment ended, there were 65,000 Turks and more than 20,000 Moroccans in Dutch industries.[12] Most migrant workers came from (rural) regions with few job prospects or opportunities for personal development and education. Seeking a better life, they migrated to Europe's industrial growth centers. These entrepreneurial workers discovered that once in the Netherlands, they had little control over their affairs and were thrown back into paternalistic support (and control) by the Dutch state and industrial companies. In practice,

here 137; Arnold van der Valk, "Ruimtelijke ordening in Nederland 1948-1973," in *Vertrouwde patronen, nieuwe dromen*, ed. Vossen, Schwegman, and Wester, 128-133, here 128.

10 Kees Schuyt and Ed Taverne, *1950: Welvaart in zwart-wit* (The Hague: Sdu Uitgevers, 2000), 199-204.

11 Company housing construction played a more modest role in Dutch postwar industrial expansion. In other places, Dinius and Vergara claim, company housing formed an integral part of postwar industrialization, especially in the Americas, where war devastations did not play a role in this period. Oliver Dinius and Angela Vergara, *Company Towns in the Americas: Landscape, Power, and Working-Class Communities* (Athens, GA: University of Georgia Press, 2011), 3.

12 "1945-heden," Vijfeeuwenmigratie, accessed June 16, 2021, https://vijfeeuwenmigratie.nl/periode/1945-heden/volledige-tekst?page=1; Bouwens et al., *Door staal gedreven*; Peet, "Werken en leren 1948-1973," 123; Vogel, *Nabije vreemden*, 83-84, 140.

this meant that immigrant workers were typically housed in temporary guesthouses and isolated "residential areas" ("woonoorden"), barracks, and hostels near factory sites—far removed from Dutch society.[13]

The government's definition of workers as temporary impacted their ability to move around. The Moluccan workers—faithful colonial soldiers of the Royal Netherlands-Indies Army and their families—were a particularly painful example of how the state treated such workers as temporary guests rather than immigrants, who should be discouraged from integrating in Dutch society. When 12,500 people from the Moluccas arrived at the ports of Amsterdam and Rotterdam in 1951 after the Dutch lost their colonial war, the loyal soldiers and their families needed housing. Local authorities allocated state camps to house unemployed Moluccans: vacant monasteries, barracks, holiday camps, and former Nazi transit camps like Westerbork, provisionally furnished. In denial about the process of decolonialization, government policy sought minimal social integration for Moluccans: the state expected that they would return to the Moluccas once the Indonesian annexation of their islands was reversed, and believed that isolated housing was the most desirable policy. By policy design, the 51 Moluccan housing sites scattered across the Netherlands, were far from urban, populated areas or were in rural peripheries, close to highways or industrial business sites. This meant the Moluccans had little contact with local Dutch people, stayed at significant distances from public services and jobs, and were prevented from starting a new life. By 1957, when the geopolitical situation in the Moluccas had not changed, the government commissioned the Verwey-Jonker Committee to investigate the Moluccans' circumstances in the Netherlands. The committee advised reversing the government's policy of temporary settlement and encouraged the integration of Moluccans by allocating housing in residential areas, also closer to jobs.[14]

Housing shortage put additional pressure on the already poor state of postwar transportation. In 1945, Hoogovens wrote to the Traffic Inspectors

13 Rita Chin, *The Guest Worker Question in Postwar Germany* (Cambridge: Cambridge University Press, 2009), 40-41; Ineke van der Valk, *Harde werkers: Migranten van het eerste uur langs Rijn en Lek, 1945-1985* (Zutphen: Walburgpers, 2009), 216, 227-228; Vogel, *Nabije vreemden*, 131. In other European countries, like West-Germany, immigrant workers from Spain, Greece, Turkey, and Yugoslavia were also housed in such isolated locations, far removed from city centers and public transport.

14 Wim Manuhutu and Henk Smeets, *Tijdelijk verblijf. De opvang van Molukkers in Nederland, 1951* (Amsterdam: De Bataafsche Leeuw, 1991); Marieke Ouweneel, *Het belang van de Molukse wijk in Nederland* (MSc Thesis, Utrecht University, 2011), 17-19; Sjoerd Post, *Anak mas – Gouden kind; Molukkers in Nederland* (Groningen: Uitgeverij Noordboek, 2001); Justus Veenman, *De sociale integratie van Molukkers* (Lelystad: Koninklijke Vermande, 1984).

about the difficulties of attracting workers from afar. The large-scale house demolition in IJmuiden by the German army prevented many Hoogovens employees from living nearby. "The majority currently live in Heemskerk, Uitgeest, and in the Zaan region. Due to the continuing shortage of labor ... we have had to provide transport for them as there are no rail connections."[15] In Eindhoven, Philips management faced similar problems. According to company magazine *Philips Koerier* in 1946: "As you know, Eindhoven and the immediate surroundings cannot meet the demand for workers. Our company would gladly see that people from other parts of the Netherlands could be accommodated in Eindhoven, but this is of course not possible given the housing deprivation."[16]

4.2 Public Transit Falls Short

The postwar decades saw a marked increase in traveled kilometers and public transit passengers. Local use of public buses and trams remained constant, around 55 million passengers a year in Amsterdam, The Hague, and Rotterdam, mostly by tram, in other cities by bus. For interlocal trips, buses and trains were the most common public mobility options, as Gijs Mom and Ruud Filarski show.[17] These numbers do not indicate for what purpose passengers used buses, trams, and trains. Some were probably commuters. Still, as I argue, when it came to accessing jobs, public transit seemed to play a limited role in meeting workers' mobility needs. Like in the interwar period, public transit failed to meet workers' needs.

Rail-based mobility was in decline in the Netherlands and elsewhere. Except in Amsterdam, The Hague, and Rotterdam, as Mom and Filarski show, tramways partially disappeared. This also happened in other European countries like Britain.[18] In neighboring Belgium too, the railroad network dwindled from 5,046 km in 1950 to 3,410 in 1998; buses replaced many of the interwar local train and tram networks.[19] The international trend, guided

15 Tata Steel Central Archives, Archive no. 4344, Inv. no. RA-05831, 2 October 1945.

16 PCA, Archive no. 642.5, Inv. no. 882, Correspondence 31 December 1948, 1, 3, and Dekker (1953), 4; "Het Personeelvervoer," *De Vrije Philips Koerier* 3, no. 8 (1946): 2.

17 Gijs Mom and Ruud Filarski, *Van Transport naar Mobiliteit: De Mobiliteitsexplosie 1895-2005* (Zutphen: Walburgpers, 2008), 263-265.

18 Colin Pooley, "Balancing Social Justice and Environmental Justice: Mobility Inequalities in Britain Since Circa 1900," in *Transport Policy: Learning Lessons From History*, edited by Colin Divall, Julian Hine and Colin Pooley (Farnham: Ashgate, 2016), 47-64.

19 Thomas Vanoutrive, "Minder inkomen, dus minder mobiel, dus minder kansen," *Armoede en Sociale Uitsluiting: Jaarboek 2018*, ed. Jill Coene et al. (Leuven/The Hague: Acco, 2018): 277-290, here 281.

by political choices favoring automobility and reducing public transit, was also reinforced in the Netherlands when employers supported bus services for their workers—as we will see in the next chapter. In the Netherlands, employers had found many of the interlocal train services instrumental in mobilizing a labor force for industry. These services were discontinued and replaced by company buses. The Dutch Hoogovens rail express rode for the last time in 1957, after decades of connecting the wider region with the blast furnaces, when more workers reverted to company buses.[20]

Dutch census data confirms this downward trend in rail commuting. It shows that in 1947, more than a fifth of workers travelled by train to jobs in other municipalities. On an aggregate level, the national share of train commuters dropped to 6 percent by 1971. In the province Noord-Holland, where interlocal commuters formed a significant share of train passengers, shares dropped less sharply than elsewhere—from 32 to 24 percent between 1947 and 1960. In the highly urbanized province Zuid-Holland, with Rotterdam as important growth center, fewer workers commuted by train and tram. Shares declined from respectively 20 percent and 18 percent in 1947, to 15 percent and 6 percent in 1960. In Overijssel, with its many textile towns, tramways were discontinued and around 17 percent of interlocal commuters kept on traveling by train. In the agribusiness lands of Brabant too, train commutes remained stable at around 10 percent. Only in Limburg, train shares increased from 5 percent in 1947 to 7 percent in 1960. But overall, the Dutch statistics bureau concluded in 1960 that rail-based mobility played only a small role in postwar commuting. Commuters typically used bicycles and mopeds for travel times up to 45 minutes (12 km by bicycle). In the period 1947-1960, "both the absolute and relative growth in the number of train commuters was proportionately less than that in the number of bus-commuters"—a trend likely encouraged by the postwar expansion of highways that literally paved the way for motorized alternatives like buses.[21]

In Rotterdam, trams often fell short of meeting workers' mobility needs. Commissioned by Rotterdam's Chambers of Commerce and Factories, the newly established Economic-Technological Institute (Economisch-Technologisch Instituut, ETI) concluded that Spaanse Polder, a newly developed industrial area on Rotterdam's northwestern fringe, was poorly unlocked by local public transport operator RET. Twenty-seven companies thought the

20 "Vaarwel Hoogovenexpres: Herinneringen bij de laatste rit," *Samen: Maandblad voor Directie en Personeel van Hoogovens en Aangesloten Bedrijven* 26, no. 10 (1957): 163.

21 Centraal Bureau voor de Statistiek, *13e Algemene volkstelling, 31 mei 1960, Deel 11: Buiten de woongemeente werkenden* (Hilversum: De Haan, 1965) H3.5, 28, SUM, 39.

Spaanse Polder's poor accessibility was a "serious handicap" in attracting and retaining local manual workers. From nearby neighborhoods Spangen, Tussendijken, and Overschie, workers walked or cycled to the Polder. But for the many unskilled workers living on the south bank of the Maas River—with its large "pool of workers" providing cheap labor—the Polder was "very difficult to reach" by public transit. A trip by tram was too expensive and too lengthy: workers had to change trams several times. ETI concluded this was a particularly serious mobility barrier for women, who were apparently less inclined than men to cycle such distances and thus relied on public trams—a national trend confirmed by the 1947 census: 52 percent of men commuted by bicycle (to jobs outside their residential municipality), versus 37 percent of women, who more often used buses, trams, and trains.[22]

Public buses did not fill the gap either. Philips management considered public transit fares too high for its workers. The services were also too limited and unreliable: workers could not get to work or arrived late for shifts. In 1948, Philips registered 1,800 weekly lost hours due to poor public transit. The coordinating body for worker bus transport in Twente's textile region agreed. Much needed men and women from towns like Losser, Overdinkel, Oldenzaal, Hengelo, Almelo, and the Achterhoek area, along with German border towns, struggled to get to factories by train and public bus. Regional public transit operators TET, TAD, GTW, and ONOG as well as Dutch Railways did not have the capacity to transport large numbers of workers during peak hours before and after day shifts. Transit operators could not get close enough to workers' departure and destination points because workers' homes and factory locations were dispersed over the wider Twente region. In addition, as the region's operators did not run services at night, workers on night shifts—becoming a more common practice to boost productivity—could not get home.[23]

Public transit was hardly ever a real option for rural and migrant workers. In the Netherlands and neighboring countries like West-Germany, migrant

22 Nationaal Archief, Archive no. 3.17.17.04, Archief Kamer van Koophandel en Fabrieken voor Rotterdam, 1922-2001, Inv. no. 1731, Economisch-Technologisch Instituut voor Zuid-Holland, "De behoefte aan personenvervoer naar en van de Spaansepolder" (1954), 1, 3-4, 6; Centraal Bureau voor de Statistiek, *12e Volkstelling annex woningtelling, 31 Mei 1947, Serie B: Voornaamste cijfers per gemeente/ Deel 6: Beroepsbevolking naar woon-, werk- en geboortegemeente* (Utrecht: De Haan, 1952), 19-21. Different RET trams brought workers to Sparta's soccer stadium near Spaanse Polder, from where a bus service provided feeder transport. Although this hub was connected to most other tram lines, for south bank workers this was still a long and expensive trip.

23 PCA, Archive no. 642.5, Inv. no. 882, Correspondence 31 December 1948, 1, 3, Dekker VIPRE Report (1953), 4; Stadsarchief Enschede, Archive no. 61, C.B.I.P.E. te Enschede, Inv. No. 2, Bedrijfsarchief, Verslag Personeelsvervoer 1955-1960, 1.

workers were typically housed in isolated locations.[24] For the Netherlands, Jaap Vogel shows in *Nabije vreemden*, how isolated these barracks were, citing Turkish worker Hasip Turan, who recalled his stay in barracks amid grasslands near Rotterdam's Botlek-Europoort industries, with hundreds of other Turks and no nearby public transport.[25] Historian Irene van der Valk notes that workers living in dire circumstances in city guesthouses in Leiden, still preferred this over moving to such isolated places because the bus connections were too poor.[26] According to the government report *Ambonezen in Nederland* (1959), Moluccan jobseekers were often unable to find (entry-level) jobs near their state-assigned, isolated residential areas. With no prior education before enlisting in the Dutch colonial army, unskilled Moluccan men depended on menial jobs and first-generation Moluccan women did not generally perform waged labor besides domestic work. Since Moluccans were largely left to their own devices on arrival in the Netherlands, with few resources to help them find employment, men sought suitable jobs in the wider area by bicycle.[27] But cycling such distances was not a structural solution for their daily commute.

Conclusion

The lack of nearby affordable housing, combined with poor public transit, posed serious challenges for workers and employers as "Mobility Barriers during Postwar Industrialization" has shown. A solution for how to get to work, a seemingly pedestrian question, was vital for workers to earn a living and for industries to upscale production. Especially for rural, cross-border, and migrant workers, not only the great spatial distance between their homes and entry-level jobs, but also the absence of affordable long-distance alternatives to cycling and public transit played a role. These workers' limited financial and social capital seemed to prevent them from resolving such challenges themselves. The government, occupied with planning transport infrastructures, appeared indecisive in addressing workers' mobility barriers. For distances up to 12 km, workers fended for themselves. They did not express their displeasure through protests against failing public transit like

24 Chin, *The Guest Worker Question*, 40-41.

25 Vogel, *Nabije vreemden*, 131.

26 Valk, *Harde werkers*, 216, 227-228.

27 *Ambonezen in Nederland: Rapport van de Commissie ingesteld bij besluit van de Minister van Maatschappelijk Werk*, (The Hague: Staatsuitgeverij, 1959), 58-59.

the unionized miners did in 1938. Rather they expressed their displeasure in their behavior by reverting to bicycles and mopeds. For longer commutes, companies intervened, removing the reasons to demand accessible transit, as we will see later. In 1948, F.E. Spat, secretary and chair of Philips' travel organization (Eigen Vervoers Organisatie), proposed that Philips should establish its own company bus services (VIPRE, founded in 1947), given the current housing shortage and mobility situation in the Netherlands.[28] It was one of the many solutions industrial employers sought to govern workers' mobility in a period of economic expansion and labor shortage.

28 PCA, Archive no. 642.5, Inv. No. 882, Correspondence 31 December 1948, 1, 3, and Dekker (1953), 4; "Het Personeelvervoer," 2.

5 Postwar Mobility Practices, 1947-1970

Amid the push for rapid postwar industrial reconstruction, workers were on the move, again. The economy was expanding, and there were plenty of jobs—though often in the wrong location. Besides jobseekers from southern Europe and northern Africa traveling by bus, train, and airplane to western Europe, the postwar decades saw increasing movements of commuting workers. In the Netherlands too, the increasing daily movements of dockers, miners, textile and factory workers in company buses was a "mass migration day after day in the Netherlands," wrote *De Tijd* reporter Gerard de Groot.[1] The Dutch Central Bureau for Statistics (CBS) judged postwar commuting ("forensisme") a "mass-scale phenomenon." And the census for 1947, 1960, and 1971 supported the assessment. By 1960, one million workers—a third of the total working population—were employed outside the place they lived, with two-thirds traveling daily to their jobs in mining, steelworks, and manufacturing.[2]

Commuting may not have been new. What did change were the distances. In the postwar era, manual workers in countries like the Netherlands and West-Germany commuted daily to industrial centers, as did workers at Italian car manufacturer FIAT, commuting up to 100 km every day.[3] Commuting had taken off in the past fifteen years, *Spil en Spoel* editors noted in 1962, particularly because "the factory bus and moped offer new opportunities for all those who could not find work in their immediate vicinity."[4] According to the Dutch 1960 census, this commuter trend resulted from "modern

1 Gerard Groot, "Massale volksverhuizing elke dag opnieuw in Nederland: Speciale busdiensten der bedrijven halen hun arbeiders van heinde en ver," *De Tijd*, 10 December 1955, 13.

2 Centraal Bureau voor de Statistiek, Centraal Bureau voor de Statistiek, *13e Algemene volkstelling, 31 mei 1960, Deel 11: Buiten de woongemeente werkenden* (Hilversum: De Haan, 1965), SUM, 38. More people commuted longer distances. The 1947, 1960, and 1971 census defined "commuter" ("pendelaar") for administrative reasons: not based on home-work travel distance but as a worker who crossed the border of their residential municipality while traveling to work.

3 Elena Dinubila, "Fight for Meaning: Representations and Work Experiences in a Greenfield Automotive Plant," *Labor History* 61, no. 1 (2020): 60-73; Alf Lüdtke and William Templer, "Polymorphous Synchrony: German Industrial Workers and the Politics of Everyday Life," *The International Review of Social History* 38, Supplement 1 (1993): 39-84.

4 "Zij komen van verre," *Spil en Spoel: Maandblad voor het Textielvak* 13, no. 3 (1962): 1.

Bek, Patrick, *No Bicycle, No Bus, No Job: The Making of Workers' Mobility in the Netherlands, 1920-1990*. Taylor & Francis Group 2022
DOI: 10.5117/9789463723183_CH05

transportation facilities." The census analysts believed the rapid development of motorized mobility alternatives to bicycles and trams expanded workers' action radius and enabled suburbanization. Since the war, "modern transport facilities have enabled persons living in a municipality with limited employment opportunities to commute to a more industrialized municipality," and middle-class people ("economically active people") to move away from cities "to areas with more pleasant living conditions," the census takers wrote.[5]

Conventional narrations of the postwar period emphasize people's commuter mobility. Transport historians Gijs Mom and Ruud Filarski characterize the 1950s and 1960s as a "mobility explosion," in terms similar to the *Spil en Spoel* editorial. They point to how the total kilometers travelled more than quadrupled in twenty years (from 19 billion in 1948 to 90 billion in 1968). Buses and cars made a difference. For example in 1950, passenger numbers in buses surpassed those in trams, peaking in 1963 with 770 million annual passengers—though public transit was gradually substituted with cars. By the late 1960s, seventy percent of the total kilometers travelled were in private cars.[6] This increasing car mileage was due to the fact that more people, including some high-paid working-class families, could afford a car. Still, new highways and having a car did not mean families used it for commuting. Male breadwinners only did this after 1970.

"Postwar Mobility Practices" nuances this "mobility explosion" narrative. First, I correct the focus on long-distance trips as new. The great majority—85 percent in 1947, 73 percent in 1960, and 66 percent in 1971—were still local blue-collar workers, who commuted shorter distances. No longer on foot, but by bicycle or moped. Second, increasing mobility cannot be solely explained by greater accessibility of new transport technologies (like cars) and accompanying state-funded infrastructures. Longer commutes may have been a middle-class lifestyle choice for some, but were more often

5 Centraal Bureau voor de Statistiek, *13e Algemene volkstelling*, SUM, 38.

6 *13e Algemene volkstelling*, SUM, 38; Gijs Mom and Ruud Filarski, *Van Transport naar Mobiliteit: De Mobiliteitsexplosie 1895-2005* (Zutphen: Walburgpers, 2008), 263-265; Hermine Molnár-in 't Veld, "De groei van het Nederlandse personenautopark," Centraal Bureau voor Statistiek, accessed June 21, 2021, https://www.cbs.nl/nl-nl/longread/statistische-trends/2019/de-groei-van-het-nederlandse-personenautopark/3-resultaten. The focus on total kilometers travelled though overrepresent longer distance trips over shorter distance trips like bicycle commuter trips, thus attributing greater value to long distance trips by motorized modes over short(er) distance bicycle trips—which within 8-12km radius still outnumbered other mobility options by 1960 this chapter shows. See for more recent data on travel distances versus time and mode: Jacqueline van Beuningen, "Automobilist het snelst op plaats bestemming," Centraal Bureau voor de Statistiek, accessed June 17, 2021, https://www.cbs.nl/nl-nl/achtergrond/2013/28/automobilist-het-snelst-op-plaats-bestemming.

a necessity for low-income workers.[7] Seeking to overcome the postwar spatial mismatch between homes and jobs, workers and employers—not the government—were key agents of change, I argue. Within 12 km distances, self-governing workers individually opted for bicycles and mopeds. For longer travel distances, employers provided company-governed buses for workers with no alternatives.

5.1 Urban and Peri-Urban Workers Keep on Cycling and Discover Mopeds

Across Europe, postwar cycling shares may have dropped. Still, cycling remained important for people's commute. *Cycling Cities* research—a historical series analyzing the development paths of cycling practices and policies in global cities—reveals a downward trend after the war (fig. 7), with a more pronounced decline outside the Netherlands. In Dutch industrial towns like Eindhoven, Enschede, Heerlen, and Kerkrade, cycling shares decreased from 80 percent in 1950 to 30 percent in 1970. These numbers, however, also indicate cycling never completely disappeared as viable mobility option.

Census analysts focused on automobility trends and commented on the declining bicycle shares in Europe, but cyclists never completely disappeared from the streets. Adri Albert de la Bruhèze, Frank Veraart, Martin Emanuel, and Ruth Oldenziel show that according to Eindhoven traffic counts, cycling made up more than half the modal split in 1947, outnumbering public transit, and cars. Cycling remained the dominant mode of travel in the electronic city Eindhoven until the 1960s. In 1961, four-fifths of Philips employees cycled to work.[8] In Rotterdam too, cyclists dominated the postwar streets—in part, fueled by rising fares and local tram operator RET's failure to connect workers' home and workplaces.[9] Similarly, for the textile town Enschede, Albert de la Bruhèze shows most commuters were cyclists, who in the late 1960s still accounted for a 64 percent modal share at main intersections of the streets Boulevard 1945 with Van Loenshofstraat, C.F. Klaarstraat, and

7 Kate Morris, "Research into travel horizons and its subsequent influence on accessibility planning and demand responsive transport strategies in Greater Manchester," paper presented at the *European Transport Conference*, Strasbourg, France (2006).

8 Frank Veraart, Martin Emanuel, and Ruth Oldenziel, "Eindhoven: Engineering a Path for Bikes?" in *Cycling Cities: The European Experience*, ed. Ruth Oldenziel et al. (Eindhoven: Foundation for the History of Technology, 2016): 53-63, here 53-54.

9 Eric Berkers et al., *Cycling Cities: The Rotterdam Experience* (Eindhoven: Foundation for the History of Technology, 2019), 23.

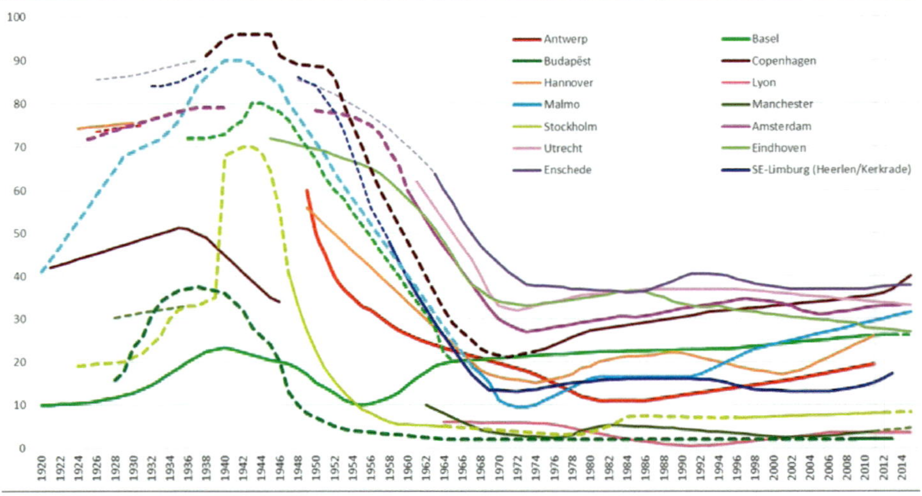

Figure 7: Trendline Cycling Shares of the Modal Split in various European cities (Source: Ruth Oldenziel et al., *Cycling Cities: The European Experience,* 2016)

Beltstraat. Over half the workers living within the city ring, thus relatively nearby, cycled to work.[10]

And there is more evidence. A 1960 photo (fig. 8) of workers leaving the factories gives a face to the thousands of men and women who relied on bicycles to get to work. "Leaving the factory" was a theme that fascinated Philips press photographers—and many other observers. Postwar photos at Philips factory gates show the jam of workers leaving after their shift. A closer look reveals workers in overalls and—typically for working men—caps, with administrative staff wearing hats and overcoats. A 1950 aerial photo (fig. 9) from the Philips press collection shows the throngs of cyclists leaving the factory at the end of their working day, capturing a common sight in industrial centers at the time.

Personnel magazines reporting on annual bicycle inspections are also evidence that many men and women employed in Dutch factories and mines continued to cycle to work. During inspections in October 1952, States Mine officers counted more than 15,000 bicycles—an indicator that bicycles played a "very important role" in commuting, according to *Nieuws van de Staatsmijnen*.[11] In 1954, local traffic police inspected 2,945 employees'

10 Adri Albert de la Bruhèze, "Enschede: An Experiment in Cycling," in *Cycling Cities: The European Experience*, ed. Oldenziel et al., 41-51, here 43.

11 "Rijwielcontrôle op onze bedrijven ontdekte ruim 4800 defecten," *Nieuws van de Staatsmijnen in Limburg* 1, no. 25 (1952): 2; "Rijwielcontrole bracht ruim 6000 defecten aan het licht," *Nieuws van de Staatsmijnen in Limburg* 2, no. 25 (1953): 3; "Een dringend beroep op alle weggebruikers," *Nieuws van de Staatsmijnen in Limburg* 4, no. 10 (1955): 2; "Rijwielcontrole op onze bedrijven,"

Figure 8: Workers crossing Eindhoven's major intersection Emmasingel-Mathildalaan by bicycle and moped, traveling home after a hard day's work. Eindhoven, 1960 (Source: Philips Company Archives)

bicycles at shipyard Wilton-Fijenoord in Schiedam and Vlaardingen, near Rotterdam, and in 1955, officers counted 1,100 bicycles at Menko and Spinnerij Roombeek in Enschede, and 19,513 bicycles at Philips factory sites.[12] The 1961 report by German traffic engineer Karl Schaechterle, who was commissioned by Eindhoven city council to map traffic flows, also showed that 80 percent of the roughly 19,000 workers scattered over Eindhoven neighborhoods Gestel, Stratum, Strijp, Tongelre, and Woensel cycled to the factories.[13]

This evidence confirms the *Cycling Cities* research with greater detail of how cycling remained a vital commuter vehicle for working-class people. And not just for short and local distances. The common assumption is that

Staatsmijnen Nieuws 12, no. 47 (1963): 2; "Controle rijwielen en brommers," *Nieuws van de Staatsmijnen in Limburg* 13, no. 46 (1964): 7.

12 "Rijwielcontrole bij Wilton-Fijenoord in de Verkeersweek," *Wilton-Fijenoord Nieuws*, 1 November 1954, 33; "Bedrijf en Maatschappij," *Mero-Meningen: Personeelsblad van N.J. Menko N.V. en N.V. Spinnerij Roombeek, Enschede* 8, no. 3 (1955): 15-16; "Bijna 10.000 rijwielen; ruim 6000 gebreken," *Philips Koerier* 12, no. 7 (1955): 1, 3.

13 Karl-Heinz Schaechterle, "Verkehrsuntersuchung N.V. Philips Eindhoven," (Ulm: Technischen Hochschule Stuttgart/ Ingenieursbüro Professor Dr. Ing. N.E. Feuchtinger, 1961), 1-6, 8-13, 22-23.

bicycles were mostly used for commuting distances under 7 km.[14] Records indicate that workers cycled even greater distances from their homes on the edge of industrial centers than assumed.[15] For many workers, bicycles (and later mopeds) were vital for accessing jobs across town and to distant industrial locations. A bicycle enabled unskilled workers to earn a living, even when hardly any job opportunities existed close to home. The 1947 census showed that over half the interlocal commuters (men in particular) within a 45-minute travel time-radius (or 12 km by bicycle) cycled to work. These statistics indicated these bicycle commuters lived just outside the administrative borders of industrial centers, in the rural periphery of places like Eindhoven. Limburg was the exception to the rule: here cycling only dominated travel distances up to 4 km, likely due to hilly terrain and (partial) company compensation for bus transport 4 km from the mines.[16]

Migrant workers cycled less. Though little is known about their mobility practices in this period, there are indications some cycled. For example, the Italians housed in hotel ship Arosa Sun could request an advance payment of 140 guilders to purchase a bicycle for the trip to the Hoogovens worksite, suggesting that short bicycle commutes were not uncommon.[17] Generally, bicycle ownership was significantly lower among migrants compared to the Dutch 1:2 bicycle ownership ratio (1965). A 1971 Dutch Foundation for Statistics report showed that in 1967-1968, bicycle ownership was two percent among Moroccan workers, 12 percent of Turkish workers, 18 percent of Spaniards, and 24 percent of Italians.[18] Bicycle ownership tells us little about usage. Low ownership figures indicate that bicycles probably did not play an important role in migrants' daily commute.

14 In 2020, the Dutch Kennisinstituut voor Mobiliteitsbeleid reported that almost a third of all commuters in the Netherlands cycle to work—most within a 5-km range from home. Matthijs de Haas and Marije Hamersma, "Fietsfeiten: nieuwe inzichten" (The Hague: Kennisinstituut voor Mobiliteitsbeleid, 2020), p. 6.

15 "Verkehrsuntersuchung N.V. Philips Eindhoven," 1-6, 8-13, 22-23.

16 This percentage is based on a study on commuters working in Amsterdam cited in the 1947 census. In Limburg, miners who did not live in the vicinity of mines, by and large came to work by two modes: bicycle or bus. Below 7km mostly bicycle, and the rest by bus. The moped range, Bijhouwer mentioned, were travel distances typically covered by commuter bus, category 1 (0-7km) and 2 (8-10km)—the 7km bicycle radius was 4km at ONM, mostly because below 4km no compensation was given for bus travel.

17 Melchior van Elteren, *Staal en Arbeid: Een sociaal-historische studie naar industriële accomodatieprocessen onder arbeiders bij het desbetreffend bedrijfsbeleid bij Hoogovens Ijmuiden, 1924-1966* (Leiden: E.J. Brill, 1986), 783, 887.

18 Nederlandse Stichting voor Statistiek, *De buitenlandse arbeider in Nederland* (The Hague: Staatsuitgeverij, 1971), 245; Jos Rietveld, "Fietsgebruik in Nederland: 1880-1996," De Oude Fiets, accessed October 28, 2021, https://www.oudefiets.nl/2-algemeen/92-fietsgebruik-in-nederland-1880-1996.

Figure 9: Aerial photo showing strings of cyclists leaving a Philips factory in the Strijp area. Eindhoven, 1950 (Source: Philips Company Archives)

In the postwar era, bicycle's share among interlocal commuters dropped to 36 percent in 1960 and 26 percent in 1971. This drop should not distract from the fact that long after the war, commuters in the Netherlands continued to cycle up to 12 km. The census bureau concluded in 1960 that despite losing "relative importance ... the bicycle ranks still first" among commuters, especially for thirty-minute or 7 km travel distances.[19] The 1960 Schaechterle report showed that a third of the workers living in Eindhoven's rural-urban periphery cycled to Philips factories.[20] By 1971, most inner-city home to work trips were by bicycle. Significantly, the 1971 census included data for the first time on commuting within city limits, showing that nationally, the bicycle made up 49 percent of city commutes.[21] Since the war in particular until the 1970s, car boosters and policymakers cast the bicycle slower and low-tech than mopeds and private cars. Yet cyclists still made up nearly a third of commutes nationally. It is evidence that as a proven, cheaper, and reliable technology, the bicycle was essential for many people getting to work.

Mopeds became important mobility options over the course of the 1950s, as workers discovered these were convenient vehicles to bridge greater distances. According to historians Mom and Filarski, Doreen Ewalds, Ger Moritz, Michel Sijstermans, and Henk-Jan Dekker, there was a close correlation between income level and moped ownership. As wages gradually increased over the 1950s, the moped became known as the car-of-the-little-man. Working-class families who could not afford a car, could buy the cheaper moped. At first, these were the breadwinning men. By 1969, of the two million mopeds registered in the Netherlands, more than half were owned by working-class families, increasingly ridden by both men and women, and 65 percent for going to school or work.[22]

Reflecting on postwar trends in commuting, the census bureau found the growing popularity of the moped or "bicycle with auxiliary engine," used by

19 Centraal Bureau voor de Statistiek, *12^e^ Volkstelling*, 11, 21, 35; *13^e^ Algemene volkstelling*, 16-17, 28; Luuk Schreven, *Documentatierapport Volkstelling 1971* (The Hague: Centraal Bureau voor de Statistiek, 2004), 175. This data illustrates a point made by historian Edgerton (2006), sociologist Shove (2012), and others about the persistence ("disappearance, partial continuity, and resurrection") of "old technologies in modern society," Citing the bicycle is a case in point. David Edgerton, *Shock of the Old: Technology and Global History Since 1900* (Oxford: Oxford University Press, 2006), xii, 45-58; Elizabeth Shove, "The Shadowy Side of Innovation: Unmaking and Sustainability," *Technology Analysis & Strategic Management* 24, no. 4 (2012): 363-375, here 363, 367, 372-373.

20 "Verkehrsuntersuchung N.V. Philips Eindhoven," 1-6, 8-13, 22-23.

21 Schreven, *Documentatierapport Volkstelling 1971*, 175.

22 Henk-Jan Dekker, "An Accident of History? How Mopeds Boosted Dutch Cycling Infrastructure (1950–1970)," *The Journal of Transport History* April (2021): 1-24; Doreen Ewalds, Ger Moritz, and Michel Sijstermans, *Bromfietsen in Nederland* (The Hague/Heerlen: Centraal Bureau voor Statistiek, 2013), 7; Mom and Filarski, *Van Transport naar Mobiliteit (II)*, 268, 270-271.

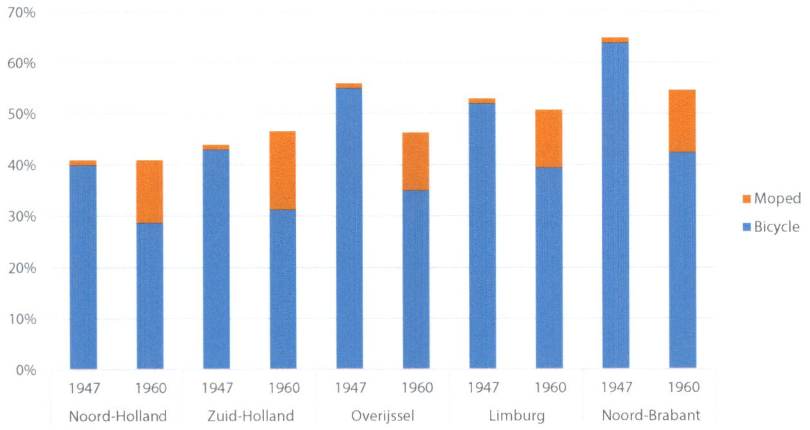

Figure 10: Modal share of bicycle and moped commuters per province (Source: Census 1947 and 1960)

13 percent—just over 100,000 workers—of all interlocal commuters in 1960, "the most striking development." The moped became a popular alternative to bicycles especially among workers living outside industrial centers. Figure 10 shows that between 1947 and 1960, the aggregate share of bicycle and moped trips decreased in provinces with major heavy industries, but moped commutes increased from virtually zero to an average of 14 percent of all interlocal commutes.[23] According to Mom and Filarski, between 1947 and 1960, cyclists substituted their bicycles for commuting by moped.[24] Census data seems to suggest this too. The question is: who switched from bicycles to mopeds and why?

Comfort and speed were important reasons to switch to mopeds. In 1960, Hoogovens personnel magazine *Samen* explained the shift from bicycle to mopeds as "Young people's desire for a motorized vehicle, the pleasant feeling of lazy satisfaction at 'not having to peddle' and the need to move faster than on an ordinary push-bike."[25] On hilly terrain, according to Frank Veraart and Manuel Stoffers, "people especially valued mopeds as a quicker, more comfortable, yet affordable alternative to the bicycle. After a hard day's work in the mine, it must have been appealing to miners to ride home effortlessly and quickly on a moped rather than exhaust themselves cycling up hills."[26]

23 *13^e^ Algemene volkstelling*, SUM 39; Schreven, *Documentatierapport Volkstelling 1971*, 175.

24 Dekker, "An Accident of History?," 1-24; Ewalds, Moritz, and Sijstermans, *Bromfietsen in Nederland*, 7; Mom and Filarski, *Van Transport naar Mobiliteit (II)*, 268, 270-271.

25 "Het verkeer en U!," *Samen: Maandblad voor Directie en Personeel van Hoogovens en Nevenbedrijven* 29, no. 5 (1960): 94-95.

26 Frank Veraart and Manuel Stoffers, "Southeast Limburg: Cycling Goes Downhill," in *Cycling Cities: The European Experience*, ed. Oldenziel et al., 65-75, here 70-71.

Contemporary observers noted mopeds had become an important commuting option—particularly for longer distances. According to ANWB magazine *De Kampioen* in 1963, mopeds were important for the economy: over two-thirds of moped riders—the "commuter army on 2 wheels"—were men between 25 and 65. For workers living in rural areas, with few mobility alternatives, moped ownership ratio was 1:5 compared to 1:7 in urban areas. "The moped has proved indispensable, especially for the daily commute involving longer distances from villages," *De Kampioen* wrote—an insight historians Mom and Filarski, Ewald, Moritz and Sijstermans, and Dekker have further detailed and analyzed.[27] Company magazines and traffic counters commented on the moped's rising popularity with workers. Virtually non-existent shortly after the Second World War, ten years later mopeds had become a common sight around factory and mining sites. Shipyard Wilton Fijenoord's bicycle parking shed housed 310 mopeds in 1954 (and 3,285 bicycles). Company magazine *Wilton-Fijenoord Nieuws* claimed that "in one year, the number of mopeds in the parking shed has tripled and is still growing."[28] In 1959, inspectors counted 1,647 mopeds at Philips in Eindhoven; and in 1963, there were more than 4,700 mopeds at the State Mines.[29] Hoogovens' Security Service noted that around 1960, moped riders almost started to outnumber cyclists at the factory gates, counting 872 bicycles and 866 mopeds. Two years later, a traffic count at the nearby sea locks showed that many more workers commuting to Hoogovens rode mopeds (1,376) than bicycles (887).[30] Comfort and speed were not the sole reasons for moped's popularity.

27 Dekker, "An Accident of History?," 1-24; L.R. van Dullemen, "Anderhalf miljoen brommers economisch nog te laag gewaardeerd," *De Kampioen* 78, no. 7 (1963): 408-409; Doreen Ewalds, Ger Moritz, and Michel Sijstersmans, *Bromfietsen in Nederland* (The Hague/Heerlen: CBS, 2013), 7; Mom and Filarski, *Van Transport naar Mobiliteit (II)*, 268, 270. Moped vs bicycle ownership was 500,000 and 5-7 million in the 1950s, and traffic counts revealed a 50-50 ratio, Dekker shows. At Rotterdam's Maas river crossings, moped riders outnumbered cyclists 40 to 60 percent. Dekker notes that mopeds trips were typically longer distances, though Rotterdam data seems to be the exception to the rule. Dekker, citing from sociologists Leendert Bak's study on the geographical dispersion of mopeds, notes that moped ownership was highest in Zeeland, Zuid-Holland, Noord-Holland, Groningen, and Drenthe—and lowest in Noord-Brabant and Limburg. Bak suggested this had to do with income levels. But in industrial cities, moped ownership was high among residents over 16-years old: Eindhoven (192), Enschede (200), Rotterdam (170).

28 B., "Rijwielstalling Vlaardingerdijk," *Wilton-Fijenoord Nieuws*, 1 November 1954, 32.

29 "Bromfietsencontrole bracht vele gebreken aan het licht," *Philips Koerier* 15, no. 18 (1959): 2; "Controle rijwielen en brommers," 3; "Rijwielcontrole op onze bedrijven," 2.

30 Tata Steel Central Archives, Archive no. 4342, Inv. No. RA-12873, Internal Correspondence 23 September 1959, Internal Memo 5 July 1961.

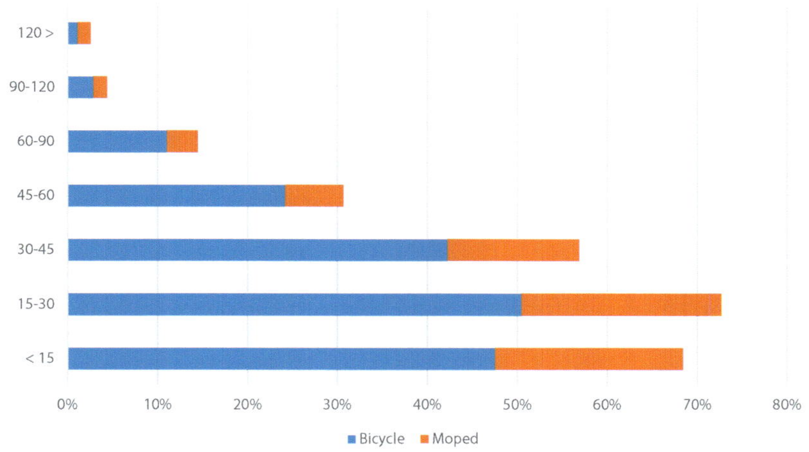

Figure 11: National bicycle versus moped commuter share per travel distances in minutes (Source: Census 1960)

Moped boosters typically argued that these expanded workers' action radius. Research, probably backed by moped manufacturers, confirmed this. In 1952, Jan Bijhouwer stated in *Journal for Public Housing and Urban Planning* (*Tijdschrift voor Volkshuisvesting en Stedebouw*) that the "maximum commuting distance" by bicycle, tram, or car was 45-minutes travel time between home and work (or nearby public transport stops). Considering the average cycling speed was 12km/h and by moped 18km/h, workers could bridge distances of 12 to 14 km in 45 minutes instead of 8 to 9 km by bicycle.[31] The Schaechterle 1961 report clocked eight percent of Eindhoven-based workers riding mopeds to work, versus eleven percent living in the city's rural surroundings within a 12 km radius: cyclists dominated the rural-urban fringe commute.[32] According to the 1960 census, mopeds took second place for the travel time range up to 45 minutes or 12 km (fig. 11). This data indicates a partial replacement of bicycle commutes, rather than a greater workers' action radius: mopeds were mostly used for travel distances predominantly covered by bicycles. Although some workers traveled distances of over an hour (about 18 km) by moped, most used them for distances under thirty minutes. Generally, mopeds thus did not seem to extend workers' travel distance, as the moped

31 Regionaal Historisch Centrum Limburg, Archive no. 17.05H, Inv. No. 46, Vervoer mijnwerkers per autobus, Correspondence 25 November 1952, and Internal Memo "Gedeeltelijke vervanging van het vervoer der arbeiders per bus en spoor door dat per bromfiets gewenst?", 21 March 1953, 2; Jan Bijhouwer, "Bromfietsbespiegelingen," *Tijdschrift voor Volkshuisvesting en Stedebouw*, 12 December 1952: 172-173.

32 "Verkehrsuntersuchung N.V. Philips Eindhoven," 8-13.

boosters claimed. They did reduce workers' commuting time for the same distances.[33] Moped usage correlated with income rather than distance.

Still, the bicycle remained important, while mopeds became popular. For workers, buses were the third favorite option. Census data indicates that most workers living within a 12-km radius from work commuted by bicycle or moped. Workers living outside industrial centers used buses.

5.2 Company Buses for Rural and Migrant Workers

Bicycles and mopeds gave workers a greater sense of autonomy over their commuting experience. Company buses, however, far less. They offered employers control over workers' mobility. Still, for workers, company buses served their needs too. Dutch rural workers, like foreign migrants, travelled greater distances to escape rural poverty and make a living elsewhere. Dwindling job opportunities in the agricultural sector, coupled with housing shortages in industrial (urban) centers, forced them to remain in their village and commute longer distances to make a living.[34] According to Langeweg in his labor study of Limburg's mining region, more workers commuted greater distances to secure work compared to the interwar period: while 67 percent still lived in their mine's town in 1930, this dropped to 57 percent three decades later.[35] In other parts of the country, longer distance commutes became more common. The company bus facilitated this practice.

The few historical studies mentioning postwar worker transport in the Netherlands indicate, but do not detail, how bus transport increasingly enabled rural jobseekers to get distant industrial jobs to escape a life of poverty and unemployment in the countryside. Unemployment in rural areas spiked in the postwar decades, as agricultural mechanization took off. In *Gezien de dreigende onrust*, labor historian Erik Nijhof discusses

33 *13^e Algemene volkstelling*, SUM 39; Schreven, *Documentatierapport Volkstelling 1971*, 175.

34 R. Peijs, "De landbouw in een stroomversnelling: Aspecten van de agrarische geschiedenis van Nederland 1945-1985," in *Vertrouwde patronen, nieuwe dromen: Nederland naar een modern industriële samenleving 1948-1973*, ed. Hélène Vossen, Marjan Schwegman, and Peter Wester (IJsselstein: VGN, 1992): 15-22, here 16-17; Molnár-in "t Veld, "De groei van het Nederlandse personenautopark." The increasing demand for foodstuffs in the Netherlands, Belgium and Germany, pushed farmers to produce more for lower prices. Mechanization helped increase productivity and enabled farmers to cut labor costs now that they needed fewer workers.

35 Serge Langeweg, *Mijnbouw en arbeidsmarkt in Nederlands-Limburg: Herkomst, werving, mobiliteit en binding van mijnwerkers tussen 1900 en 1965* (Hilversum: Verloren, 2011), 64, 67. In Eygelshoven, circa 4 percent and in Geleen around a third of the miners travelled more than 15 km to the mines.

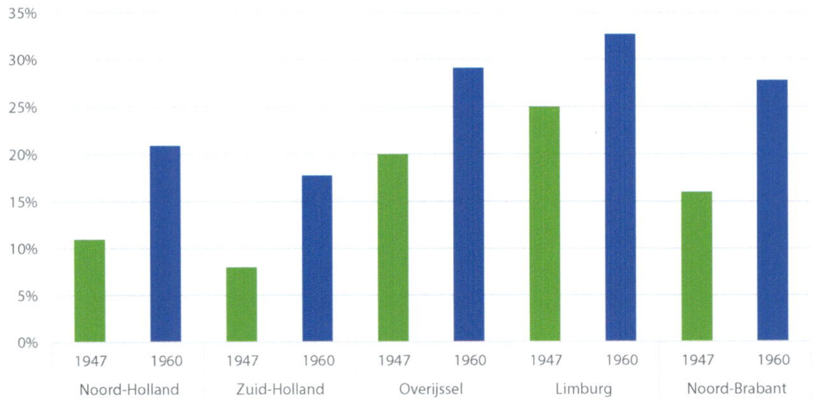

Figure 12: Provincial modal share of bus commuters (Source: 1947 and 1960 census)

poor, uneducated rural workers traveling by bus to Rotterdam's port.[36] The availability of long-distance bus services served as a valve for regional unemployment, as social historian Jan Zwemer argues in *Zeeland 1950-1965*. Men from Tholen and Sint-Philipsland, northern peninsulas of the province Zeeland, with the highest postwar unemployment rates in the country, commuted by small and large buses to Rotterdam and other distant industries.[37] Historical sources confirm this trend.

The bus was particularly popular with male rural workers. The 1947 and 1960 census reports showed that commuters used the bus more in the provinces with large industrial sites. The number of bus commuters tripled to 185,000 workers, with a relative increase of 14 to 25 percent of all interlocal commutes. With travel times of 45 minutes (15 km by bicycle), the bus outweighed all other modes, attaining it highest share (68 percent) for commuting distances of two hours or more.[38] The greater the distance, the higher the share of bus commuters.

Neither the 1947 nor the 1960 census report distinguishes between public buses and chartered or company buses. Considering the decline in public transit passengers, this increase in bus commuters was likely the result of the rapidly expanding company bus system. Moreover, the long distances workers travelled on these buses indicate the services operated across

36 Erik Nijhof, *'Gezien de dreigende onrust in de haven': De ontwikkeling van de arbeidsverhoudingen in de Rotterdamse haven 1945-1965* (Amsterdam: IISG, 1988), 138-143.

37 Jan Zwemer, "De pendelarbeiders van Tholen en Sint-Philipsland," in *Zeeland 1950-1965*, ed. Jan Zwemer (Vlissingen: Uitgever Den Boer/De Ruiter, 2005): 401-428, 401-402, 418-419.

38 *13e Algemene volkstelling*, SUM, 39.

provincial and national borders. These were outside public transit concession areas that bound operators to designated routes.

According to state traffic inspector-general Th.M.B. van Marle in 1953, despite government efforts to bring the transport sector back into shape, it was Dutch businesses rather than the state that successfully mobilized workers from far and wide by bus in the first postwar decade. Company managers, by cooperating with local bus operators and setting up company-owned bus services, succeeded in collectively transporting thousands of workers on "an unprecedented scale," he wrote, expanding from nearly twelve and a half million passengers in improvised commuter buses in 1946, to almost 31 million in the company-orchestrated bus system by 1950.[39] Catholic weekly *De Tijd* journalist Groot reported in 1955, for example, how large Dutch companies like the mines in Limburg, electronics company Philips, and shipyard Wilton-Fijenoord ran daily bus services. Smaller companies did likewise. A cake factory in Zaandam had three buses to pick up workers 60 km away. A company in Alkmaar with about 300 employees did the same, just like smaller industries in Nijmegen.[40] These company buses enabled workers from outside industrial centers, in places with no (public) mobility options, to access factory and mining jobs.

Many long-distance bus commuters lived in rural towns and villages near industrial centers, but also further afield. Schaechterle reported that of the 34,000 Philips employees working in Eindhoven, 12,000 lived outside the city, of which two-fifths commuted by bus. In 1959, many Philips workers came from villages like Best (10 km) and Geldrop (7 km), but also from much further away in Zeeland province (more than 100 km from Eindhoven). Around a fifth lived in villages like Duizel and Eersel (17 km) and the Belgian Campine region (50 km), while another 15 percent lived in the countryside around towns like Roermond (60 km).[41] Thousands came from far and wide to work in Twente's textile industry. In 1962, textile magazine *Spil en Spoel* reported that Twente's workers came from rural places like Gramsbergen, Staphorst, Rühlermoor in Germany, and Barger-Compascuum (50-80 km), and even from the further away province Drenthe or West German border-regions (160 km). According to the magazine, by 1962 "the bus has become an indelible symbol of our textile companies."[42]

39 Th. M. B. van Marle, *Het binnenlands vervoer na de bevrijding: 1944/45-1952* (The Hague: Nederlands Verkeersinstituut, 1953), 44, 49-51.

40 Groot, "Massale volksverhuizing," 13.

41 "Verkehrsuntersuchung N.V. Philips Eindhoven," 1-6, 8-13, 22-23.

42 "Zij komen van verre." 1.

Scholars at the time examined commuters' social background. In his 1963 study of dock workers in Amsterdam and Rotterdam, sociologist Pieter ter Hoeven identified the rural commuters as "a category of workers, varied in composition." They were profoundly different from city residents because they came from rural areas without industrial bases, plagued by poverty and unemployed due to rapidly mechanizing agriculture. Compared to their urban counterparts, rural workers were often younger than 45, lower skilled, and had not completed primary school education, Ter Hoeven found. Used to physically demanding agricultural work and seasonal wage-shifts, these rural commuters were generally more willing to accept labor-intensive work. Urban workers often switched jobs if they disliked the working conditions, but rural commuters had "an all-dominating orientation towards [higher] wages," according to Ter Hoeven.[43] Other scholars, then and now, confirm his findings, including sociologists Lambooy and Mak (commissioned by Hoogovens) and E.F. Bosch (1957), as well as historians Zwemer in *Zeeland 1950-1965* (2005) and Langeweg in *Mijnbouw en arbeidsmarkt in Nederlands-Limburg* (2011).[44]

Most bus commuters were Dutch citizens. Some were male migrant workers, traveling from their guesthouses or remote housing quarters by company buses. Other migrant workers lived in isolated quarters at walking distance from industrial sites, like the on-site Hoogovens hotel ship Arosa Sun for Italian and Spanish workers. Generally, housing locations were further afield and only accessible by company buses. Melchior van Elteren notes how Hoogovens had buses running between its blast furnace locations in IJmuiden and Spanish living quarters like Casa del Norte in Heerhugowaard throughout the 1960s.[45] Van der Valk mentions that Turkish workers in Leiden's isolated living quarters were picked up by small buses every day.[46] In Limburg, Italian and Hungarian migrant workers were transported by bus from their "bachelor dwellings" to the mines (fig. 13), according to national newspaper *Trouw* (1957); and *Schiedamsche Courant* (1971) wrote that Wilton-Fijenoord's Turkish workers housed in a former retirement home in

43 Pieter ter Hoeven, *De Havenarbeiders van Amsterdam en Rotterdam: Een sociologische analyse van een arbeidsmarkt* (Rotterdam: Stenfert Kroese, 1963), 115-117, 126-127.

44 E.F. Bosch, "De sociale problematiek van Tholen en Sint-Philipsland, Deel 1: Het Forensisme," (Middelburg: Stichting Zeeland voor Maatschappelijk en Cultureel Werk, 1957); J.G. Lambooy and J.H. Mak, "De pendelaar in woon-, reis-, en werkomgeving," (IJmuiden: Koninklijke Nederlandsche Hoogovens en Staalfabrieken, n.d.); Langeweg, *Mijnbouw en arbeidsmarkt*, 63-72; N.C. Schouten, "Pendelarbeiders en industrievestiging op Tholen," (Middelburg: Stichting Zeeland voor Maatschappelijk en Cultureel Werk, 1961); Zwemer, "De pendelarbeiders," 401-428.

45 Elteren, *Staal en Arbeid*, 783, 887.

46 Ineke van der Valk, *Harde werkers: Migranten van het eerste uur langs Rijn en Lek, 1945-1985* (Zutphen: Walburgpers, 2009), 216, 227-228.

Figure 13: Italian miners working at Limburg's State Mines travelled to their guesthouses on mining company buses. Simpelveld, 1951 (Source: Regionaal Historisch Centrum Limburg, Fotocollectie DSM, PR6456, Creative Commons: CC-BY-SA)

Poortugaal, south-west of Rotterdam (10 km) went to work in small company buses.[47] Moluccans—formally not migrant workers—were also picked up by buses from their isolated barracks (fig. 14), according to a 1959 government report.[48] For them too, waged work was worth the long commute.

This was also true for cross-border workers. According to *Philips Koerier* (1946), Belgian women (fig. 15) who travelled long distances by bus from the rural Campine area preferred to work at Philips because of the higher wages in industry: "You can buy anything in Belgium, but it's expensive, and folk don't have much money, so they come to Philips, ... to work, earn money."[49] From the Belgian Campine region, with no industrial center, young men and women "were swarming into the Netherlands," *Het Parool* wrote in 1949. Asked why she crossed the Dutch border, one "Marjanneke" replied, "You can't earn much in our area."[50] Newspapers citing census data showed cross-border commuting

47 "Bus met mijnwerkers op vrachtauto," *Trouw*, 7 November 1957, 1; "Honderd Turken in voormalig bejaardentehuis," *Schiedamsche Courant*, 13 January 1971, 3.

48 *Ambonezen in Nederland: Rapport van de Commissie ingesteld bij besluit van de Minister van Maatschappelijk Werk* (The Hague: Staatsuitgeverij, 1959), 58-59.

49 "Susanneke werkt in Nederland," *De Vrije Philips Koerier* 3, no. 7 (1946): 2.

50 Philips Company Archives, Archive no. 642.5, Inv. no. 882, newspaper clipping "Dagelijks volksverhuizingen van België naar hier: Lange rijen autobussen voeren fabriekspersoneel aan," *Het Parool*, 26 October 1949.

Figure 14: Moluccans from the Dutch Indies residing at Schattenberg settlement wait for the bus to take them to work. Often devoid of nearby and regular public transport, Moluccans could only access distant jobs by company bus. Westerbork, 1951-1960 (Source: Drents Museum/ Moluks Historisch Museum)

increased in the postwar period. The number of Belgian workers employed in the Netherlands' southern provinces Noord-Brabant, Limburg, and Zeeland, had grown to 17,417 in 1964, and 24,645 in 1970. Crossing borders in the other direction, Dutch workers commuting daily increased too (from 37,000 to 52,500 between 1963 and 1971), mostly to Belgian and West-German construction industries. Most cross-border Dutch commuters lived in Limburg and worked in West-Germany, where wages were still between 18 and 28 percent higher than Dutch wages. According to historian Sophie Bouwens, few commuters in Southeast Limburg used public transit—cross-border connections were generally poor or lacking—and travelled by company-chartered vans or touring cars. For better wages, they accepted the longer travel time.[51]

Company buses also had a downside. Living further away from industrial jobs affected rural and migrant workers' lives in ways not experienced by

51 Katholiek Documentatie Centrum, Archive no. 370, Nederlands Katholiek Vakverbond (NKV), 1894-2001, Inventory no. 26718, Stukken betreffende arbeidspendel, newspaper clipping, G.W. Groeneveld, "Omvang en oorzaken van de grens-pendel," *Gelderlander*, 14 October 1972; Sophie Bouwens, *Over de streep: Grensarbeid vanuit Zuid-Limburg naar Duitsland, 1959-2001*, vol. 71, Maaslandse Monografieën (Hilversum: Verloren, 2011), 118-119, 146. Around 1970 cross-border commuting by car had become common too.

workers staying within or just outside industrial centers. According to sociologists Ter Hoeven and Lambooy and Mak, company buses were not voluntary but a "forced form" of commuting: the uneven geographic distribution of (industrial) jobs forced especially lower-income and lower-skilled workers to commute greater distances to make a living. Lambooy and Mak also observed that Hoogovens' bus commuters generally belonged to the "lowest category of workers," even though agricultural workers, who were skilled in their own sector, were designated by the company as "entirely unskilled" upon entry. Companies considered them as belonging to the "working class, with a lower standard of living and social status as a result of ... on the one hand the social and economic situation in their municipality, and ... on the other hand the social and economic situation in their working area."[52] For these reasons, they were usually assigned easily replaceable functions with little prospect of promotion.[53] In this sense, the increased bus mobility did not just offer new possibilities, but also represented a form of deskilling.

Ununionized migrant and rural workers, who depended on their employer for their commute, found their position more precarious than local workers. For one, Zwemer mentions that commuters from rural Zeeland were typically the first to be fired in Rotterdam's port when company results were disappointing. They were, despite lower wages, simply more expensive than local workers due to higher transport costs.[54] At Hoogovens, according to Lambooy and Mak, rural commuters were not hired for key positions and usually first in line for dismissal given the cost of commuting and uncertainty associated with longer commuting distances.[55] The government report on Moluccan workers' socioeconomic position raised similar concerns. With declining labor demand, the cost of commuting and the uncertainty about arrival times likely motivated employers to fire Moluccan bus commuters first.[56] Although long distance commuting enabled workers to land distant, better paid jobs, at the same time, commuting also reinforced their precarious labor position.

Migrant and rural workers also had more difficulty connecting with a local union and workers' communities and thus fewer joined the unions. At Hoogovens, the union of highly skilled metal workers did not have chapters in remote villages, forcing rural commuters to attend meetings in Beverwijk near Hoogovens—prohibitive given the travel distance and fixed bus timetables.

52 "De pendelaar," 2-4, 7, 36, VI 22.

53 A. ter Hoeven, "De positie van de pendelarbeider op de locale arbeidsmarkt," *Sociologische Gids* 6, no. 1 (1959): 17-30; Nijhof, *Gezien de dreigende onrust*, 138-143; Zwemer, "De pendelarbeiders," 414.

54 "De sociale problematiek," 8-9, 20-21, 31-32, 41, 44; Zwemer, "De pendelarbeiders," 406, 408.

55 "De pendelaar," 36, VI 22, VII 39-11, Appendix 38.

56 *Ambonezen in Nederland*, 58-59.

Figure 15: Philips' own company VIPRE buses lined up at Beukenlaan, waiting for Belgian women to board after their factory shift. Eindhoven, 1949 (Source: Philips Company Archives)

Just like migrant workers, rural commuters had no outside party to voice complaints about working conditions or renumeration but their department heads. Migrant workers and rural commuters typically worked separately from local, higher-skilled workers. Immigrants, who struggled to be part of local communities, often faced cultural barriers to connect with their host society. The fact they were often housed in isolated locations with fellow countrymen, far from services and recreation, did not help. Dutch rural workers struggled too with social isolation.[57] In 1965, Paul van Beckum, a worker at the coal conveyor belt at Hoogovens loading dock, told regional newspaper *Verenigde Noordhollandse Dagbladen* that taking part in Hoogovens' social activities was in practice impossible: "It is too far to attend regularly."[58] A 1959 government report noted that Moluccan workers were unable to bond with other (Dutch) workers after their working day, because "the work bus is waiting" at the end of their shift.[59] As we have seen, the isolation was policy design.

57 "De pendelaar," 36, VI 22.

58 Paul van Beckum, "Ik pendel," *Verenigde Noordhollandse Dagbladen: Bijlage Kerstmis* (December 1965): 30-31.

59 *Ambonezen in Nederland*, 58-59.

Conclusion

In contrast with the previous decade, workers and employers, not the state, were key players in shaping the postwar everyday commute in the Netherlands. In their "mobility explosion" theses, mobility historians Mom and Filarski emphasize the increasing number of travelled kilometers, thereby highlighting long distance travel and (private) motorization, obscuring shorter distance trips made on foot, bicycle or moped. "Postwar Mobility Practices" brings to the fore that most workers lived in the urban and peri-urban surroundings of industrial jobs, and commuted distances up to 12 km. The bicycle and later also the moped were indispensable for these workers' everyday commute. For workers living further afield, we have seen that public transit fell short. Company buses were the primary mobility alternative at the time.

The chapter shows that while scholars' idea of a postwar "mobility explosion" suggests a sudden outburst with velocity and force, we should also consider the fragmented, uncontrollable aspects of mobility. The central explanation in this scholarship is the availability of new (motorized) transport technologies for more people and centrally planned transport infrastructures, enabling movement from A to B. We have, however, seen that shaping mobility is a social process which owes much to the agency of individuals and collectives who negotiated their commute according to their interests and abilities. The following two chapters examine the social forces and processes underlying this explosive growth of commuters and travelled kilometers. They focus on how through politics of care and control, employers tried to secure workers' mobility in the postwar period.

6 Disciplining Cyclists and Moped Riders

In 1961, H.M.J. van de Wall was driving his car across Hoogovens site. As safety editor of company magazine Together (*Samen*), he regularly reflected on work-floor and traffic safety. Approaching an on-site intersection with a maximum speed limit of 30 km/hour, he slowed down for a cyclist who had to give him right of way. The cyclist did not stop, causing a situation that the editor said illustrated the appalling lack of cycling workers' traffic knowledge. He honked his horn, and as the cyclist just skimmed past his bumper, shouted: "Were you sleeping, brother?" The man got off his bicycle and stepped closer. Fear struck Van de Wall. He would not have been surprised if the cyclist had grabbed a hammer from his back pocket. The cyclist, however, calmy replied, "I am wide awake. Why?" Van de Wall explained the situation. But instead of thanking him for this insight, "asking what I thought of a cigar or whether I rather preferred shag tobacco," the cyclist wrongly claimed that, coming from the right, he (always) had right-of-way.[1]

The editor's eloquently phrased outburst oozed class bias. To Van de Wall, the cyclist's likely preference for shag tobacco over a more dignified cigar, and fear of a hammer in the cyclist's back pocket, made him a mere unskilled worker, unlike himself, an upper middle-class car driver. Van de Wall claimed he shrugged off the incident. Still, he found it "very scary" the cyclist was convinced that coming from the right gave him priority over all traffic, "thus continually risking his life, the more so because of his strong and stubborn conviction about being on the 'right' path (for how long?). This could cost the fellow sooner or later his life."[2] Van de Wall articulated a high-class hierarchization of road users. The world was divided between upper middle-class car owners, representing reason, knowledgeable about traffic regulations, and gentlemanly behavior, versus working-class cyclists, framed as indifferent, unruly, and unfamiliar with traffic regulations. The distinction has a history.

The class-based distinction between high-class car drivers and working-class cyclists was a recurrent theme of ongoing (class) struggles over road space

1 H.M.J. van de Wall, "Over veiligheid gesproken .... Recht(s) is Recht(s)," *Samen: Maandblad voor Directie en Personeel van Hoogovens en Nevenbedrijven* 30, no. 9 (1961): 178.

2 "Over veiligheid gesproken .... Recht(s) is Recht(s)," 178.

Bek, Patrick, *No Bicycle, No Bus, No Job: The Making of Workers' Mobility in the Netherlands, 1920-1990*. Taylor & Francis Group 2022
DOI: 10.5117/9789463723183_CH06

across the United States and Europe since the interwar period, as historians Peter Norton, Adri Albert de la Bruhèze, Ruth Oldenziel, Martin Emanuel, and others have shown.[3] Walking and cycling became more hazardous as the speed of car traffic increased after the interwar period. As velocity, efficiency, and flow became the guiding road design principles, engineers and traffic associations developed strategies to govern these practices. Jennifer Bonham claims that road users who resisted the "ordering of the speedy street" by cycling in their own way were considered "irrational" since they simultaneously resisted the "order of safety."[4] We can trace how the safety discourse that framed cyclists as irrational emerged in postwar company magazines.

For manual workers, bicycles and mopeds were indispensable for accessing jobs in the first postwar decades. Governments and employers, however, considered that cyclists and moped riders literally rode in the way of (future) automobility. This traffic safety discourse, as Norton has shown for the United States, was informed by car-centered notions, favoring automobility over other modes.[5] This also applied to Europe and the Netherlands, Oldenziel and Albert de la Bruhèze show.[6] "Disciplining Cyclists and Moped Riders" expands on their observations. Besides state and engineering efforts to segregate cyclists and moped riders from main roads via separate bicycle paths, I show how employers were also key in disciplining cycling and moped riding workers. This chapter argues that with employers' involvement, the car-centered safety discourse intertwined with a company discourse of paternalistic care and control over workers. Complementing labor history, I show that company control extended beyond the factory gates.

To solve the explosion of traffic accidents, companies and national safety organizations rode in tandem to instigate a new class-based traffic regime, extending control over dockers, miners, textile and factory workers also outside the work site. Preventing traffic accidents ultimately served another goal: maintaining a healthy workforce to ensure productivity. The following shows how companies' fears about traffic safety intertwined with their aims

3 Martin Emanuel, "Stockholm: Where Public Transit Eclipses Cycling," in *Cycling Cities: The European Experience*, ed. Ruth Oldenziel et al. (Eindhoven: Foundation for the History of Technology, 2016), 149-159; Peter Norton, "Jaywalking and the Invention of the Motor Age Street," *Technology and Culture* 48, no. 2 (2007): 331-359; *Fighting Traffic: The Dawn of the Motor Age in the American City* (Cambridge, MA: MIT Press, 2008); Ruth Oldenziel and Adri Albert de la Bruhèze, "Contested Spaces: Bicycle Lanes in Urban Europe, 1900-1995," *Transfers* 1, no. 2 (2011): 29-49.

4 Jennifer Bonham, "Transport: Disciplining the Body That Travels," *The Sociological Review* 54, no. 1 (2006): 55-74, here 64-65.

5 Peter Norton, "Four Paradigms: Traffic Safety in the Twentieth-Century United States," *Technology and Culture* 56, no. 2 (2015): 319-334.

6 Oldenziel and Albert de la Bruhèze, "Contested Spaces," 29-49.

to control workers' mobility and safeguard productivity. They instituted rules and norms—and deployed company-specific strategies to discipline working-class road users beyond the factory gates.

The organizing force in safety campaigns was the Dutch Alliance for Safe Traffic (Verbond voor Veilig Verkeer, VVV). Established in 1932 by car drivers responding to the public outcry about declining road safety during the early onset of motorized traffic—the safety alliance set up dozens of local branches across the country to promote traffic safety via lobbying and educational campaigns for schoolchildren and workers—and was part of an international movement. Constantly seeking ways and means to convey its safe traffic message, the organization formed alliances with other social actors. In 1958, the VVV's special Committee "Companies and Safety" published a booklet *Enterprise and Traffic Safety* (*De onderneming en de verkeersveiligheid*) to urge employers to collaborate in making workers more "traffic-minded."[7] In the 1950s, industrial employers were an important ally in disciplining and educating workers who commuted on foot, bicycle and moped.

6.1 Companies Fear for Workers' Safety in Postwar Traffic

Van de Wall's class-based view on car drivers' right of way over cyclists had been long in the making. Both non-state and state actors had been addressing the rising rate of road accidents and deaths since the interwar period, but most safety campaigns focused on curbing pedestrians' and cyclists' movements to favor an unhampered flow for motorists. Crash records show that authorities framed workers' cycling and moped practices as responsible for their injuries and fatalities. What kind of mechanisms did industrial-capitalist companies create to discipline workers?

First, the employers' disciplining efforts took place around and beyond the factory gate. Workers entering the worksite had to dismount from their bicycle or moped on passing the gatehouse and then continue on foot. On-site traffic lights, road signs, and markings directed workers.[8] At Hoogovens, officials issued speeding moped riders with "written warnings" and fines,

7 Verbond voor Veilig Verkeer, *De Onderneming en de Verkeersveiligheid* (Utrecht: Verbond voor Veilig Verkeer, 1958), 11.

8 B., "Rijwielstalling Vlaardingerdijk," *Wilton-Fijenoord Nieuws*, 1 November 1954, 32; "Verkeerssituate bij Staatsmijn Hendrik vraagt uiterste voorzichtigheid," *Nieuws van de Staatsmijnen in Limburg* 4, no. 10 (1955): 2; "Er wordt gebouwd," *Mero-Meningen: Personeelsblad van N.J. Menko N.V. en N.V. Spinnerij Roombeek, Enschede* 9, no. 10 (1957): 9; Van de Wall, "Over veiligheid gesproken .... Wijzer geworden," *Samen: Maandblad voor Directie en Personeel van Hoogovens*

measuring on-site speed with a "gatsometer," even "banning people for a certain time from traveling round the site with their motorized vehicle," according to *Samen* in 1960.[9] These on-site traffic rules were formalized in factory regulations. In 1953, the shipyard *Wilton-Fijenoord Nieuws* reported that it was "forbidden, without permission in writing, to have a bicycle outside the designated parking shed" except for red and white painted "company bicycles." Such prescribed workers' parking places ensured unhampered traffic flows.[10] The Traffic Brigade at Philips removed wrongly parked bicycles and mopeds scattered over the Eindhoven sites. Workers at Enschede's textile factories had to request a "bicycle parking space" from the wages department.[11]

Outside the factory gates, workers also faced employers' disciplinary measures. In general, company intervention had focused on workers' mobility options, transport modes, and supporting infrastructures. Employers also believed workers' behavior needed calibration to ensure they arrived safely and on time—and thus productivity. Controlling workers' movements outside the gates became a proxy for safeguarding capitalist production.

Company managers worried about workers' traffic safety, not only for fear of injury and loss of life. They also were concerned about increasing costs. The national safety organization VVV established a committee for businesses (Bedrijven en Verkeersveiligheid) with directors of companies like Philips and Hoogovens. In 1958, the committee declared "The red light is burning!" to emphasize unsafe traffic had reached a critical point and action was needed. Based on statistics, the committee observed an "alarming rise" in road accidents, tripling in over twenty years (between 1935 and 1957 from 43,000 to 134,000 accidents: 800 to 1,700 traffic fatalities and 14,000 to 39,000 injured).[12] The dramatic increase meant loss of working days and higher company expenses for

en Nevenbedrijven 27, no. 2 (1958): 31; "Groen, geel, rood," *Samen: Maandblad voor Directie en Personeel van Hoogovens en Nevenbedrijven* 30, no. 1 (1961): 17.

9 "25 kilometer en niet harder," *Samen: Maandblad voor Directie en Personeel van Hoogovens en Nevenbedrijven* 29, no. 7 (1960): 132.

10 "Officiële Medelingen: Sociale Rubriek," *Wilton-Fijenoord Nieuws*, 1 June 1953, 2.

11 "Ter beteugeling van de verkeersonveiligheid: Beperkende voorschriften voor parkeren op complex Strijp te Eindhoven," *Philips Koerier* 14, no. 9 (1957): 8.

12 Verbond, *De Onderneming en de Verkeersveiligheid*, 5. Costs for Dutch society were estimated at 88 million guilders in 1948, 235 million in 1955, or one percent of the gross national income. The committee based their findings and recommendations and experiences from participating companies, shared by Ir K. Kooij (director Chamotte Unie, Geldermalsen), F.W.E. Spies (director Hoogovens), Drs B.M. Sweers (secretary of management AKU, Arnhem), A.H. Verkade (director Verkade Fabrieken, Zaandam), Mr F.E. Vlielander Hein (director N.V. Hollandse Draad- en Kabelfabriek, Amsterdam), and Ir D.A.C. Zoethout (director Philips, Eindhoven), and Mr N.B. ten Bokkel Huinink and Jhr Mr J. van Doorn of VVV.

Figure 16: Police officers inspecting the rear light on a Philips employee's bicycle. Eindhoven, 1955 (Source: Philips Koerier/ Company Archives)

sick leave, they insisted. According to the traffic safety organization, employers had to think of traffic safety as a proxy for protecting/controlling the labor force, essential to industrial-capitalist production and economic growth. Besides the tragic loss of workers for their families, production loss caused huge damage to Dutch society and industries: temporary replacement, recruiting and familiarizing new employees, the cost of social unrest, and other consequences of an employee's illness or death. Moreover, compared to work-floor accidents, the duration of workers' absence was significantly higher with traffic incidents, the organization insisted.[13] The argument resonated.

Company managers blamed absenteeism and the loss of productivity and capital on traffic accidents. According to *Philips Koerier*, "One million eight hundred thousand workdays were lost in 1954."[14] Due to the 884 employees injured in road accidents, 11,441 workdays (13 days for each accident) were lost. Accident-related absenteeism cost the community and society the tidy sum of 88 million guilders in 1948, not counting lost workdays, according to the textile industry's *Spil en Spoel*.[15] By 1955, the number of unworked

13 *De Onderneming en de Verkeersveiligheid*, 5-6.

14 "Heen en weer: Veel ongevallen tijdens spitsuren," *Philips Koerier* 15, no. 6 (1958): 2.

15 "Zet het licht op groen," *Spil en Spoel: Maandblad voor het Textielvak* 9, no. 3 (1958): 36-37.

days reached a staggering 1.8 million, the editors warned. In 1954, another textile magazine *Mero-Meningen* reported about twenty-seven workers involved in collisions with their bicycle or moped. The magazine calculated that 378 working days were lost and about two thousand guilders on injury compensation—not to mention production loss. According to statistics from the Health Service's Bandage Room, cited in *Mero-Meningen* of 1955, 500 traffic-related sick days at Menko and Spinnerij Roombeek, cost 4000 guilders in injury compensation.[16] The business press often blamed cyclists and moped riders for these increasing traffic injuries and believed the solution was to guide and curtail their movements.

Employers also blamed poor maintenance. An important tool to ensure proper standards was the annual bicycle and moped inspection. The VVV urged employers to carry out such inspections in cooperation with the police and ensure defects were repaired as soon as possible, preferably in company repair shops during working hours.[17] Employers complied, and these annual inspections (fig. 16) became common practice across industries.

"BICYCLE OWNERS-EMPLOYEES OF W.F., PAY ATTENTION!

Who's that ringing my bell? You are probably thinking. Listen.

Do you readers know that there was a time, the police were not held in too high esteem by us? This was, by the way, a long time ago. Now it's all different. We also see these people as our educators. They are now going to do us a favor, as it were. Because, what's going to happen? As part of the traffic safety week in our company, from 30 October to 6 November, the Schiedam traffic police will examine your bicycle for defects on Tuesday 2 November, **free of charge**. It is therefore essential that you do not lock your bicycle that day, so that the brakes can also be tested. You will receive a label on your bicycle showing any defects. In return for this free inspection, you can have your bicycle repaired or do it

16 "Bedrijf en Maatschappij," *Mero-Meningen: Personeelsblad van N.J. Menko N.V. en N.V. Spinnerij Roombeek, Enschede* 8, no. 3 (1955): 15-16; "Ongevallen omlaag," *Mero-Meningen: Personeelsblad van N.J. Menko N.V. en N.V. Spinnerij Roombeek, Enschede* 8, no. 13 (1956): 3.

17 "De veilige fiets," *Nieuws van de Staatsmijnen in Limburg* 6, no. 13 (1957): 8; "Is uw rijwiel wel in orde?" *Philips Koerier* 15, no. 2 (1958): 2; Verbond, *De Onderneming en de Verkeersveiligheid*, 9-10. Time and time again, managers and local traffic police officers reminded workers to ensure their bicycles and mopeds were up to standard. In 1957, State Mine Maurits even placed a temporary display "the safe bicycle" where people collected their weekly pay ("loonhal"). The two-week exhibition presented a bicycle in proper condition and listed the most common defects found during the 1956 and 1957 inspections, with the motto: "Check your bike today, tomorrow could be too late."

yourself, as long as it gets done. Otherwise, we would have learned nothing from this traffic safety campaign.

So Tuesday, 2 November, do not lock your bicycle.

You need not fear theft; there is extra security.

Thanks in advance."

Bicycle inspection notification for Wilton-Fijenoord Employees, 1954 (Source: F.S, "Rijwielbezitters-medewerkers van W.F., let op!," *Wilton-Fijenoord Nieuws*, 1 October 1954, 24.)

In October 1954, dockworkers found a note on their parked bicycles, also published in *Wilton-Feijnoord Nieuws*. Employees were asked to not lock their bicycles in the parking sheds, enabling local police officials to conduct a safety inspection.[18]

Around annual inspection time, companies claimed that workers often rode poorly maintained bicycles and mopeds. Tire shortages may have been resolved by the late 1940s, but until 1963, workers had little disposable income to spend on repairs or buy new tires due to the government's stringent wage policies. Workers rode second-hand bicycles or older models, sometimes in poor condition—not out of preference, but necessity. Employers put the onus on workers. In 1955, bicycles inspected at Philips factories displayed numerous defects: "The millions of cyclists weaving their way day in day out through increasingly complex and hazardous traffic are causing needless danger by not sorting these defects."[19] That year *Nieuws van de Staatsmijnen* raised the same concerns because miners' bicycles revealed "so staggeringly many defects." Eight years later, little had changed: inspections still identified about 3000 shortcomings in 8000 miners' bicycles and 600 in 5000 mopeds. By October 1952, the problems in 15,000 bicycles had increased from 4800 to 6100 in just a year's time. Most common defects involved unpainted rear fenders (2300), rear lights not working (1700), missing head and rear lights; malfunctioning brakes and bicycle bells; and worn-out saddles, pedals, and carrier straps. Officers notified all offenders that they risked a fine if failing to fix the defects. The 1955 State Mine inspection of 17,000 bicycles and mopeds revealed no

18 F.S, "Rijwielbezitters-medewerkers van W.F., let op!," *Wilton-Fijenoord Nieuws*, 1 October 1954, 24.

19 "Bijna 10.000 rijwielen; ruim 6000 gebreken," *Philips Koerier* 12, no. 7 (1955): 1-3, here 1, 3.

fewer than 4500 shortcomings. More than a quarter of the vehicles did not meet traffic safety standards. In November 1961, the State Mines inspected 8100 bicycles and 4600 mopeds for "roadworthiness" ("rijvaardigheid.") In the bicycles, they detected 3000 anomalies, 550 in mopeds.[20]

Inspectors created many lists to prove their point. Inspections in October 1954 of nearly 3000 bicycles and 250 mopeds at shipyard Wilton Fijenoord during "Traffic Week" reported: 67 malfunctioning brakes; 813 broken or missing bells; 165 defective head lamps; 395 defective rear lights; 1400 non or "badly painted mudguards". And on the mopeds: 32 faulty brakes; 35 broken or missing bells; 38 poorly or unpainted mudguards. The annual inspection in 1955 at Enschede's textile factories found 1100 bicycles had over 540 similar defects. Officers at Philips noted "1289 defective bells, 198 faulty brakes, 516 head lamps not working, no fewer than 1050 broken rear lights, 2729 not white enough fenders, and 302 bicycles were in an overall poor condition."[21] At Philips, nearly 10,000 bicycles had more than 6000 defects. "Considering that an estimated total of 21,000 bicycles are parked at Philips, this means that over 13,000 cyclists or moped riders are going around in traffic with vehicles that are partially unroadworthy." Two teams concluded that 65 percent of employees were using vehicles with faults that could endanger traffic safety every day.[22] The fact that so many workers chose to ride allegedly poor bicycles, was likely due to too low incomes for expensive new bicycles or repairs. This also raises doubts about whether these dangers were real or projected, as workers seemed to have no problem riding their bicycles and mopeds for distances up 12 km.

Local police fined workers if they failed to follow up on the repairs. If workers could not fix their bicycle, companies offered bicycle repair services at modest rates, usually deducted from wages. The Oranje Nassau Mines had on-site repairmen on the payroll and Philips provided on-site repair shops on working days between 7.15 am and 1 pm, and 2.30 to 6 pm. For the "unlucky ones" needing repairs outside these times, there was a toolbox so that cyclists and moped riders could do repairs themselves. Textile mills Menko and Spinnerij Roombeek also offered to paint their employees'

20 "Rijwielcontrole bracht ruim 6000 defecten aan het licht," *Nieuws van de Staatsmijnen in Limburg* 2, no. 25 (1953): 3; "Een dringend beroep op alle weggebruikers," *Nieuws van de Staatsmijnen in Limburg* 4, no. 10 (1955): 2; "Rijwielcontrole op onze bedrijven," *Staatsmijnen Nieuws* 12, no. 47 (1963): 2; "Controle rijwielen en brommers," *Nieuws van de Staatsmijnen in Limburg* 13, no. 46 (1964): 7.

21 "Bijna 10.000 rijwielen; ruim 6000 gebreken," 1, 3.

22 "Rijwielcontrole bij Wilton-Fijenoord in de Verkeersweek," *Wilton-Fijenoord Nieuws*, 1 November 1954, 33; "Bedrijf en Maatschappij (8/3)," 15-16.

fender white at the factory for 15 cents.[23] All to ensure employees rode proper bicycles and mopeds.

Authorities also blamed workers' disobedient behavior. Eindhoven traffic police claimed in 1955, that workers were renowned for their unruly behavior *en route* to work, "the audacious way these guys speed along the road."[24] Employers believed unruly behavior and lack of traffic knowledge were the norm among many workers who commuted by bicycle and moped. According to *Mero-Meningen*, men and women using the roads to Menko and Roombeek "are undisciplined and unwilling to work together, they have no sense of propriety in traffic."[25] Despite calling on all road users including car drivers to heed safety, company magazines singled out cyclists and moped riders: slow(er) road users like cyclists should adjust to motorized traffic rather than the other way around. According to J.M. Buitelaar, editor of mining magazine Coal (*Steenkool*) in 1951, Dutch workers traveling on foot, bicycle, and moped were "rude" and had "brutality." Whereas a Dutchman was commonly regarded as "charming, pious," Buitelaar declared "in getting ahead, he transforms into a complete heathen." The editor did not shy away from the most graphic term to assign blame: the cyclist "justifies his speed even under the hands of a surgeon, who is gathering the crushed limbs, that is, only if the victim's head is still intact."[26] This unruly behavior was most noticeable when everyone poured out of the factories into the chaos at the end of a shift. Timmermans, head of traffic police, warned dockers in *Wilton-Fijenoord Nieuws* the same year: "how many defy road safety, how many seek the traffic chaos and prefer the hospital to their card night, cozy living room, or watching football. They ride as if they are the only ones on the road."[27] Cyclists seemed to enjoy their individual freedom. Company magazines, however, felt workers were responsible for the unsafe situations outside the factory gates.

Companies sought to police workers' traffic behavior. In 1957, *Philips Koerier* surveyed police officers at six Philips locations about the five most common "mistakes" cycling workers made. According to this survey,

23 Regionaal Historisch Centrum Limburg, Archive no. 17.05H, Inv. no. H49; "Jaarverslag over 1955 van de Veiligheidscommissie van N.J. Menko N.V. en N.V. Spinnerij Roombeek," *Mero-Meningen: Personeelsblad van N.J. Menko N.V. en N.V. Spinnerij Roombeek, Enschede* 8, no. 11 (1956): 9; "Openstelling rijwielreparatiewerkplaats," *Staatsmijnen Nieuws* 16, no. 16 (1967): 10.

24 "Bijna 10.000 rijwielen; ruim 6000 gebreken," 1, 3.

25 "Bedrijf en Maatschappij (8/1)," 3-4.

26 J.M. Buitelaar, "De mens en het monster," *Steenkool: Bedrijfstijdschrift van de Nederlandse Steenkolenmijnen* 6, no. 22 (1951): 506-507.

27 J.B. Timmerman, "Verkeersveiligheid," *Wilton-Fijenoord Nieuws*, 1 October 1951, 2-3.

"the majority of cyclists in the Netherlands—six and a half million of the 10,899,521 residents who have a bicycle—commit roughly the same minor sins." The editors were convinced that most of their 60,000 readers were aware of committing offences like failing to indicate directions or give right of way, riding too far right or more than two side by side, jumping the queue at traffic lights, and malfunctioning bicycle lights. Workers might still have assumed that like before, the road was a shared rather than a segregated space. Philips, however, emphasized that not roads and local traffic circumstances were the problem, but "the people who constitute the traffic." The magazine went on: "The mistakes boil down to carelessness, nonchalance, alleged haste, laziness, insufficient knowledge of the rules." Philips traffic expert H.F. Hanegraaf, former chief of police, agreed. He singled out young cyclists as "brats" who hindered car drivers "by swerving back and forth" in front of car bumpers, showing off, and thinking "the road is as much mine as the car driver on his comfy seat."[28] In 1958, a traffic safety committee also concluded that Dutch cyclists and pedestrians—many of them workers—were "not very disciplined," clearly seen from the perspective of drivers whose cars were more difficult to maneuver and thus thrived in segregated spaces and with strict regulations.[29] In 1959, Hoogovens warned its workers that in traffic, "some drivers are taken over by an evil spirit, and it seems like they are waging a Cold War against each other." The "sad lists of dead and injured" were the result of "human shortcomings and errors that caused the majority (80 percent) of accidents," safety journalist Van de Wall wrote. "We know many of these mistakes already: lack of responsibility ... driving too fast, cutting in, reckless."[30]

28 "Geen richting aangeven: de meest gemaakte fout van wielrijders," *Philips Koerier* 13, no. 16 (1957): 1; "Fouten tegen verkeersorde bevorderen onveiligheid," *Philips Koerier* 13, no. 16 (1957): 3. Philips had factories in Baarn, Eindhoven, Hilversum, Roermond, Roosendaal, and Zwolle. Cycling scholar Marco te Brömmelstroet argues that cyclists swarm like starlings and car drivers like geese, according unambiguous and clear rules, leaving little room for improvisation. Ignoring the uniqueness of bicycle mobility, traffic planners and engineers typically apply the "goose logic" to governing cyclists in traffic. Marco te Brömmelstroet, "Een andere kijk op fietsen: Over stofzuigers, ganzenlogica en spreeuwenzwermen," Cycling Academics, accessed August 30, 2021, http://cyclingacademics.blogspot.com/2014/06/een-andere-kijk-op-fietsen-over.html.

29 Verbond, *De Onderneming en de Verkeersveiligheid*.

30 "Het verkeer en U!," *Samen: Maandblad voor Directie en Personeel van Hoogovens en Nevenbedrijven* 28, no. 11 (1959): 210-211; H.M.J. van de Wall, "Over veiligheid gesproken .... Leerzaam," *Samen: Maandblad voor Directie en Personeel van Hoogovens en Nevenbedrijven* 29, no. 2 (1960): 34.

Employers accused workers of lacking basic traffic etiquette. Company magazines in the 1950s believed working-class cyclists and moped riders had an appalling lack of understanding of traffic regulations and were indifferent about learning them. Traffic education—or learning how to make way for car traffic, as Norton has pointed out—only became part of the school curriculum in 1959. Most unskilled workers had not had formal traffic training, while middle and upper-class car drivers learned the basic traffic regulations to get their driver's license. According to Philips, the many accidents that occurred with cyclists in Eindhoven "clearly demonstrate that they don't know everything yet."[31] Hoogovens agreed.[32] So did mining operators. According to *Nieuws van de Staatsmijnen*, this ignorance posed an everyday risk to "thousands of Dutch people who do not really realize the life-threatening dangers facing them and so nonchalantly take this risk, just like a child playing with a razor."[33] The discourse surrounding traffic safety in company magazines was alarmist. It emphasized the real—and imagined—threat of traffic incidents.

The traffic safety discourse was also patronizing and reductionist regarding bicycle and moped riders' street behavior and (working-class) riders' knowledge about the rules of the road. It articulated class bias against working-class people: only through policing and traffic education could working-class cyclists and moped riders become fully-fledged road users—and citizens—instead of liabilities for industrial productivity. Accidents led to high social costs—in particular for businesses. The decade showed increasing car traffic and higher speeds. Companies, together with the car drivers' safety organization VVV explained traffic incidents by blaming employees' unruly behavior in traffic and ignorance of the rules. They did not frame the cause of accidents in terms of drivers with more wealth and access to power increasing their motorized speed. Instead, they assigned the blame to the more vulnerable working-class riders, who were often captive cyclists with limited alternatives: workers had to adjust to motorized traffic and change their behavior rather than vice-versa. Safety campaigns focused on workers who had to be mindful in traffic, keep their vehicles in proper condition, and avoid getting in the way of other (faster, motorized) traffic. Policing and educational campaigns were the two main tools.

31 "Cursus voor bromfietsers," *Philips Koerier* 13, no. 3 (1956): 2.

32 "Het verkeer en U!," *Samen: Maandblad voor Directie en Personeel van Hoogovens en Nevenbedrijven* 29, no. 9 (1960): 170-171.

33 Buitelaar, "De mens en het monster," 506-507.

6.2 Policing and Schooling

Van de Wall's belief that cycling workers needed to be educated was a sign of the times. Authorities and employers believed that knowing and obeying traffic regulations were prerequisites not just for becoming a road user, but also a modern, hard-working citizen. Companies invested significantly in educating workers about traffic regulations and promoting self-discipline. Employers acknowledged that the men and women living within a certain travel radius of the factory or mining sites were left to their own devices—discouraged or prohibited from using company buses, as we have seen, and rode bicycles or mopeds to work. Workers' behavior, however, still needed assimilation to new traffic norms to prevent them falling victim. Nationally operating car-oriented VVV in collaboration with employers sought to change workers' traffic behavior through policing (law enforcement and engineering inventions) and education. Employers played a key role.

First were regulations and enforcements. In 1958, VVV recommended employers implement traffic regulations on their premises like speed limits, road signs and markings, traffic lights, and in particular, "rules for entering and leaving sites." Funneling these movements ensured safe and unhampered on-site traffic by implementing one-way-traffic, placing "arrows, signs etc. ... routing traffic .... To parking places, directing to correct bicycle parking areas." Prohibiting workers to park and ride bicycles and mopeds on factory sites would also benefit order and safety, the association insisted.[34] Such disciplinary measures did reduce traffic accidents on factory sites. Philips recorded 173 on-site traffic accidents in 1955. Three years later the number was down to 109. Hoogovens Medical Service reported that after new speed limits in September 1957, the average twenty-nine accidents with injuries a month dropped to nine.[35] Funneling workers' movements obviously worked, judging by the fewer on-site incidents. The same could not be said outside the factory gates.

The level of control companies could exercise was limited outside because they had no (official) mandate beyond the factory gates. Still, companies cooperated closely with the police and VVV to jointly implement a repertoire of intervention and discipline. The interventions focused on employees' behavioral change and self-discipline rather than top-down policing. Editors of the *Wilton-Fijenoord Nieuws* special 1951 safety edition wrote that work floor accidents and

34 Verbond, *De Onderneming en de Verkeersveiligheid*, 7-8.

35 "Ongevallencijfer daalt," *Philips Koerier* 13, no. 26 (1957): 3; "Sociale Luisterpost: Minder ongevallen," *Philips Koerier* 15, no. 18 (1959): 2; Wall, "Wijzer geworden," 31.

on-site safety issues fell under the jurisdiction of company safety committees and traffic brigades: "The only solution here is self-discipline."[36] On-site workers' bodies were disciplined, outside the gates it was workers' minds.

Dutch companies attempted to inculcate workers' minds through educational drives about traffic and individual responsibility, following an international trend.[37] The VVV together with industrial employers rolled out traffic safety campaigns, initially targeting schoolchildren. In the postwar period, they singled out industrial workers because they made up most road users. In 1948, the *Wilton-Fijenoord Nieuws* characterized those who were inconsiderate towards other road users "anti-social." The magazine stated: "A large company had no need of people without a sense of community." The editor reminded readers: "In our W.F. community, we depend on each other. We must learn to get on with each other, work together with as little friction and conflict as possible. And we have to abide by certain rules created thanks to years of experience and agreements made with a certain goal in mind." Employees had to know their responsibility vis-a-vis their colleagues, also in traffic.[38] "Othering" of so-called unruly road users (pedestrians and cyclists) had been common in the interwar period. The difference: after the war, it was taken up by industrial employers too, who argued unruly road users did not belong in a company's work force. Companies urged workers to demonstrate middle-class, civilized, "gentleman-like" behavior in traffic. According to the shipyard news: "Traffic accidents are exactly the same as in companies, ... people had to prevent 75 percent of the accidents; 15 percent can be avoided through protection etc., so therefore 10 percent remained unavoidable. It is the same with traffic accidents, but the figure indicating how many people have to prevent them is much higher: they say 94 percent." Schiedam police inspector Kloeseman wrote to employees in 1954 in *Wilton-Fijenoord Nieuws* that "Self-discipline is better than imposed discipline." R. Schoepp, chair of Philips' Committee for Daily Affairs (Commissie voor Dagelijkse Zaken, CDZ) also found self-discipline inside and outside the factory premises the key.[39]

36 Directie, "Ter introductie," *Wilton-Fijenoord Nieuws*, 1 October 1951, 2.

37 According to Bonham, the Anglo-Saxon safety campaign "Look Both Ways" in schools and media. Bonham and Cox argue that safety was "woven into discourses on the nature of public good and of socio-economically responsible behaviour, reinforcing the linkage between travel behaviours and 'responsible citizens.'" Bonham, "Transport: Disciplining the Body That Travels," 67-68; Jennifer Bonham and Peter Cox, "The Disruptive Traveller? A Foucauldian Analysis of Cycleways," *Road and Transport Research* 19, no. 2 (2010): 42-53.

38 S, "Veilig verkeer in en buiten het bedrijf," *Wilton-Fijenoord Nieuws*, 1 October 1948, 9.

39 W.F. Kloeseman, "Wilton in het verkeer," *Wilton-Fijenoord Nieuws*, 1 October 1954, 21-22.

These campaigns typically appealed for gentleman-like behavior that editor Van de Wall evoked in his confrontation with the cyclist. In 1958, Dutch company magazines promoted "Neighborly love also in traffic," demanding from workers a great deal of "self-discipline and courtesy in traffic." As the VVV urged: "We will have to fight this mentality: create awareness that coping with traffic will be easier for all concerned if road users treat one another with courtesy, kindness, and respect each other's issues."[40] In the following decade, company magazines repeated endlessly the principles of self-discipline and gentlemanly behavior in traffic, hinting at working-class readers. In 1962, editor E.R.S. Krudop of company magazine *Mero-Meningen* emphasized the "gentleman" credo expected of both upper-class employees and unskilled workers. "Gentle" referred to upper-class merits like "friendly, observant, helpful, attentive, fair, proud, merciful, generous." These were personal characteristics not only acquired by birth but could be cultivated through "educating hearts and minds." He defined a 'gentleman-like' person as "someone who does their ultimate best in all life situations not to hurt another person's feelings. ... He puts himself in the shoes of his interlocutors (even his opponents)." This was "naturally a challenge for everyone to work towards such ideals," also in traffic.[41] Editor Van de Wall cited Amy Groskamp-ten Have's influential etiquette publication ("Hoe hóórt het eigenlijk?"), as "useful ... for people aiming ('from the cradle to the grave') to give as little as possible offence." He believed her lessons also applied to people's interactions in traffic, especially self-discipline.[42]

The educational campaign could turn personal. Shipyard employees involved in a traffic accident received a personal letter from their Safety Service: "Reports show, that on (such and such a date) you incurred an injury in a traffic accident." As victim, the employee was now expected to be aware of the risks in modern traffic, appealing to their "responsibility as a civilized human being" and their "sense of reality."[43] Companies rolled out yet more formal educational campaigns promoting desirable behavior.

Traffic education became institutionalized nationally. By 1959, traffic education was part of the curriculum at primary schools. Next, the VVV

40 "Zet het licht op groen," 36-37; "Ook in het verkeer naastenliefde?," *Spil en Spoel: Maandblad voor het Textielvak* 16, no. 5 (1965): 6-7; Verbond, *De Onderneming en de Verkeersveiligheid*, 11, 13.

41 E.R.S. Krudop, "Wees een Heer in het verkeer," *Mero-Meningen: Personeelsblad van N.J. Menko N.V. en N.V. Spinnerij Roombeek, Enschede* 15, no. 8 (1962): 3-6.

42 Van de Wall, "Over veiligheid gesproken .... Hoe hóórt het eigenlijk?," *Samen: Maandblad voor Directie en Personeel van Hoogovens en Nevenbedrijven* 35, no. 1 (1966): 17.

43 Verbond, *De Onderneming en de Verkeersveiligheid*, Bijlage: Persoonlijke brief zoals in gebruik bij de Dok- en Werf-Maatschappij Wilton-Fijenoord N.V. te Schiedam.

recommended managers educate employees by organizing competitions in safe riding; traffic courses with an exam and certificate; propaganda evenings or a safe traffic stint during personnel events, inviting speakers, showing safety films and performing short sketches; hanging posters in factory areas, visible to all employees; distributing safe traffic stickers, providing VVV-magazine *Wegwijs* and other safety booklets along with employees' pay packets, and publishing articles.[44] *Philips Koerier* advertised VVV traffic safety courses for the men and women working at Philips, who lived in or around Eindhoven. The course with about 300 registered participants consisted of four weekly evening classes from 7.30 to 9.30 pm at Police headquarters, which also provided instructors. Philips subsidized the course: its employees only paid one guilder for the full course instead of 2.50 for VVV members or 3.50 for non-members. At four evening classes, workers watched films on traffic safety and were instructed in regulation and signs, given a technical explanation of various types of road vehicles, and discussed specific traffic situations using drawings and photos.[45] Similarly, Enschede's traffic police and the VVV organized a course for car drivers, cyclists, and moped riders, costing 2.50 guilders. Bearing in mind the work shifts, the courses were held in company canteens, with three or four 2-hour sessions, sometimes singling out certain road users such as moped riders to instruct them in traffic regulations specific to this group."[46]

Migrant workers also attended courses—most likely at their managers' insistence. With the help of translators, Spanish and Turkish workers were instructed by local police about traffic regulations and road signs near their company living quarters. *Algemeen Dagblad* wrote in 1964 that in Hengelo, after Turkish workers were involved in traffic incidents with bicycles, mopeds, motorcycles, and older American cars, local police wanted to put a stop to this by providing courses with vivid slideshows.[47] According to *Het Vrije Volk* (1965) police officers visualized right of way to Turkish

44 "Eindelijk een échte Schriftelijke Verkeerscursus," *Mero-Meningen: Personeelsblad van N.J. Menko N.V. en N.V. Spinnerij Roombeek, Enschede* 9, no. 8 (1957): 8; "Zet het licht op groen," 36-37; Marco te Brömmelstroet, "Mobility Language Matters," (Amsterdam: De Correspondent Uitgevers, 2020), 48; Verbond, *De Onderneming en de Verkeersveiligheid*, 11-12; Van de Wall, "Over veiligheid gesproken .... Alle-overschaduwend," *Samen: Maandblad voor Directie en Personeel van Hoogovens en Nevenbedrijven* 30, no. 4 (1961): 83.

45 "Cursus voor bromfietsers," 2; "Verkeerscursussen voor personeel in Eindhoven," *Philips Koerier* 15, no. 5 (1958): 5.

46 "Veilig Verkeer," *Mero-Meningen: Personeelsblad van N.J. Menko N.V. en N.V. Spinnerij Roombeek, Enschede* 8, no. 6 (1955): 10 ; "Veilig Verkeer," *Mero-Meningen: Personeelsblad van N.J. Menko N.V. en N.V. Spinnerij Roombeek, Enschede* 14, no. 5 (1960): 2.

47 "Verkeersles voor Turken," *Algemeen Dagblad*, 19 September 1964, 1.

workers using scale models of intersections.[48] Vlaardingen traffic police told *Algemeen Dagblad* in 1972 that many rural Turkish workers had no experience with busy traffic or cycling—thus explaining the relatively high number involved in traffic accidents that year, despite being a local minority group.[49]

Companies used their magazines as conduit to convey traffic safety messages to employees. These played an important role in educating workers about healthy lifestyle, worldly affairs, and leisure time, but also proved important channels for VVV campaigns and traffic safety discourse. Via magazines, companies crafted an 'imagined traffic community' and schooled workers about desired behavior while commuting by bicycle or moped. Insisting their core policy was to instill a sense of community, corporations prided themselves on developing a wide range of social facilities, as propagated in their magazines. This process of crafting an imagined traffic community also singled out cyclists and moped riders as unruly: they threatened the community idea. In company towns, there was no place for "a-socials."[50]

Companies developed an educational repertoire through columns, advertisements, cartoons, and quizzes to raise awareness about traffic (un)safety in their working-class readers. In free magazines, VVV published articles and advertisements (less frequently than tourist organization ANWB). Workers in companies across industrial sectors and regions were bombarded with messages like "keep to the right, make way. Stay in the lane that is assigned to you," and "Showing understanding in his driving style for what we urgently need in traffic: safety and flow."[51] Throughout 1956, *Nieuws van de Staatsmijnen* ran a recurring column, "The Cyclists

48 "'Kijk, dit is nou een onverharde weg ...' Turken worden ingewijd in de gevaren van het verkeer," *Het Vrije Volk*, 13 August 1965, 3.

49 "Verkeersles voor Turken," *Algemeen Dagblad*, 2 November 1972, 11.

50 Political scientist Benedict Anderson, on the social construction of nation-states, famously argued that "imagined communities" are "cultural artefacts of a particular kind ...," as "a complex 'crossing' of discrete historical forces; but that, once created, they became 'modular', capable of being transplanted, with varying degrees of self-consciousness, to a great variety of social terrains, to merge and be merged with a correspondingly wide variety of political and ideological constellations." He considered all political communities, "larger than primordial villages of face-to-face contact," to be imagined "because the members of even the smallest nation will never know most of their fellow-members, meet them, or even hear from them, yet in the minds of each lives the image of their communion." Benedict Anderson, *Imagined Communities: Reflections of the Origin and Spread of Nationalism* (London: Verso, 1983), 48-50.

51 Verbond voor Veilig Verkeer, "Snel – sneller – snelst," *Spil en Spoel: Maandblad voor het Textielvak* 9, no. 11 (1959): 18-19; "Snel ... sneller ... snelst," *Mero-Meningen* 11, no. 11 (1959): 12; "Snel – Sneller – Snelst," *De Wekker* 13, no. 2 (1959): 3; "Houd rechts – Maak ruime," *De Wekker* 15, no. 1 (1961): 7-8; "Hangoren: Uw plaats op de weg," *De Wekker* 17, no. 2 (1963): 6.

in Traffic," discussing the importance of heeding traffic regulations, and recognizing traffic signs, and always reminding readers to ride on the right side of the road.[52] Hoogovens columns had a fixed format. Employees were asked to take a short test and tick the right box—the correct answers were published in the next *Samen* edition.[53] A 1950 "Safety sign puzzle" (fig. 17) illustrated unsafe traffic where workers failed to wear helmets or take necessary precautions. Unruly cyclists in traffic featured in this puzzle too: cycling without holding the handlebars, a cyclist being pulled along by a car, and riding three side-by-side.[54]

Cartoons did the same. Shipyard Wilton Fijenoord conveyed theirs via a company mascot. In its October 1957 issue, "Victor Safety" ("Victor Veiligheid") gave the workers commuting to and from the shipyard every day, the following rhymes:

> "Watch out when you're walking or on your bike.
> Watch out for the safety of all of us.
> Look to your left and to your right, look ahead and behind your back.
> Those who don't watch out, are stubborn and foolish.
> In the towns and in the countryside, look to all sides."[55]

The cartoons (fig. 18) in a 1957 *Philips Koerier* article left no doubt that road users were to blame for the many unsafe traffic situations in Eindhoven. Each portrayed cyclists as ignorant, unruly, and reckless: three cyclists side-by-side obstructing cars, not indicating directions, not giving right of way, and not even sparing pedestrians on the sidewalk. In contrast, other road users like car drivers—and pedestrians—were portrayed as utterly frustrated or startled by the improper behavior of cyclists, to whom they fell victim.[56] In short, Van de Wall's class-based outburst against the working-class cyclists he encountered while driving was well embedded in employers' discourse and practice. It reflected employers' attempt to discipline workers' mobility outside the factory gate.

52 "De wielrijder in het verkeer," *Nieuws van de Staatsmijnen in Limburg* 5, no. 11 (1956): 3.

53 "Het verkeer en U! (28/11)," 210-211; "Het verkeer en U! (29/2)," 30-31; "Het verkeer en U! (29/9)," 170-171.

54 "Oplossing Veiligheidstekenpuzzle," *Samen: Maandblad voor Directie en Personeel van Hoogovens en Nevenbedrijven* 19, no. 2 (1950): 18-19.

55 "Veiligheid," *Wilton-Fijenoord Nieuws* 20 (1957): 17-18. ("Kijk uit wanneer je loopt of rijdt./ Kijk uit voor ons aller veiligheid/ Kijk naar links en rechts, voor en achteruit / Wie niet uitkijkt is eigenwijs en dom./ In stad en land, kijk toch uit naar alle kant.")

56 "Fouten tegen verkeersorde bevorderen onveiligheid," 3.

Figure 17: Hoogovens "Safety Sign Puzzle" in personnel magazine *Samen* (1950), Source: *Samen*/ Tata Steel Central Archives.

13
14-15
18-21
16
9
1
BRAM BLOM
NOV 1949

Figure 18: *Philips Koerier* (1957) cartoon, educating its cycling readers in traffic regulations: "do not ride more than two side by side" (Source: *Philips Koerier*/ Philips Company Archives)

Conclusion

Post-World War II employers intervened in workers' mobility to gain and retain access to cheap labor. Cyclists and moped riders were policed and instructed how to ride safely to work and avoid obstructing motorized traffic on their commute. These interventions were strategies to cost-efficiently care for and control workers' commutes to and from steel companies, docks, mines, and textile factories, inscribed by power relations and industrial paternalism. Underlying this framing of working-class riders as unruly and ignorant was a process of "othering" that had taken root already in the interwar period, as mobility scholars have pointed out. The burden of responsibility for traffic incidents was placed on unruly bodies and undisciplined minds. Across industrializing nations, non-motorized road users were increasingly subjected to stigmatizing language as motorized traffic gradually dominated the streetscapes.

As "Disciplining Cyclists and Moped Riders" demonstrates, employers expressed their concerns with a veil of caring, but their desire to control was strong too. Blaming and framing workers as unruly and ignorant road users indicated the ambition to control workers' movements rather than a sense of compassion. Cyclists and moped riders were divided into hierarchical—inferior and superior—categories, and exposed to strategies intent on assimilating them to the norm set by those in power. Companies' use of stigmatizing, frightening language to convey their message about traffic safety was instrumental. Their instructions justified disciplining workers' behavior.

Employers had no mandate outside the factory gates to top-down discipline workers. Therefore, they collaborated with others to orchestrate educational campaigns. Via annual bicycle and moped inspections, traffic education, and company magazines, workers were reminded time and again about the need for safety standards, self-discipline, and gentleman-like behavior in traffic. Safety alliance VVV and companies envisioned that crafting an ideal community of workers would help (re)socialize and create shared moral values. The rationale behind these efforts was that knowing one's place in traffic reduced traffic incidents. If everyone complied, that would lead to safer commutes and reduce production loss. Being educated about traffic and behaving accordingly served a common goal: reducing traffic incidents—benefitting oneself, the company, and society at large. Like in other countries, the Dutch traffic safety association was the driving force behind these efforts. "Disciplining Cyclists and Moped Riders" revealed that industrial companies proved key agents too.

7 Mobilizing Rural and Migrant Workers by Company Bus

Amid postwar industrial growth, the company bus enabled employers to expand labor markets, as labor historians have mentioned without addressing the underlying politics.[1] We have seen that company buses enabled rural and migrant workers in places with no mobility alternatives to access farther afield jobs. With these buses their travel horizons expanded considerably. This chapter shows these buses also allowed employers to control how and when mostly unskilled and ununionized workers travelled. Historians have shown that the postwar era was marked by far-reaching rationalization and mechanization of production to control the work process. I argue that postwar company buses also should be seen as an extension of these paternalist-capitalist power relations.

The twentieth-century mechanization of production and application of regulatory systems—formal and informal—enabled employers to prescribe when, where, and how workers performed their tasks on the shopfloor. The interwar rationalization of industrial production found even wider application in the Netherlands during postwar reconstruction and industrialization compared to previous decades.[2] Company buses, like other social welfare amenities that employers provided, were presented as an in-kind remuneration

1 Bram Bouwens et al., *Door staal gedreven: Van Hoogovens tot Tata Steel, 1918-2018* (Bussum: Uitgeverij Thoth, 2018), 119-120; Serge Langeweg, *Mijnbouw en arbeidsmarkt in Nederlands-Limburg: Herkomst, werving, mobiliteit en binding van mijnwerkers tussen 1900 en 1965* (Hilversum: Verloren, 2011), 63-72; Erik Nijhof, *'Gezien de dreigende onrust in de haven': De ontwikkeling van de arbeidsverhoudingen in de Rotterdamse haven 1945-1965* (Amsterdam: IISG, 1988), 138-142; Jan Zwemer, "De pendelarbeiders van Tholen en Sint-Philipsland," in *Zeeland 1950-1965*, ed. Jan Zwemer (Vlissingen: Uitgever Den Boer/De Ruiter, 2005): 401-428.

2 Marcel van der Linden, "Work Incentives and Forms of Supervision," in *Handbook Global History of Work*, ed. Karin Hofmeester and Marcel van der Linden (Berlin/Boston: De Gruyter Oldenourg, 2018): 469-489, here 479-481; Johan Schot, Harry Lintsen, and Arie Rip, "Betwiste modernisering," in *Techniek in Nederland in de twintigste eeuw. Deel 1: Techniek in ontwikkeling, waterstaat, kantoor en informatietechnologie*, ed. Johan Schot, et al. (Eindhoven/Zutphen: Foundation for the History of Technology /Walburgpers, 1998), 19-36, here, 20.

Bek, Patrick, *No Bicycle, No Bus, No Job: The Making of Workers' Mobility in the Netherlands, 1920-1990*. Taylor & Francis Group 2022
DOI: 10.5117/9789463723183_CH07

for manual workers, a form of company care, even across national borders. Just like government-subsidized public welfare, however, company welfare was overladen with moral values, (de)legitimizing certain forms of living, working, and, as the chapter shows, forms of mobility. In their mobility austerity measures of the 1940s, the state and employers had determined whose mobility should be safeguarded (unskilled workers in industries critical to the war effort). These measures were guided by capitalist notions about who produces value in society and, consequently, whose mobility access should be assisted. Guiding the organization of workers' mobility was inseparable from postwar industrial expansion and the continued shaping of people's bodies as productive bodies—their movements constituting productivity.

For commuter buses, there were different regulations for each aspect of travel. Routes and timetables imposed a strict time-discipline over workers' movements, as well as travel behavior getting on and off buses, and en route. It was strictly regulated who could or could not get on the buses. The successful, undisturbed flow of these meticulously planned commuter buses depended on excluding some workers over others—in this, pricing schemes and personalized bus cards were instrumental.

7.1 Employers as Driving Force Behind Worker Buses

As we have seen, employers took an active role in organizing transport for their workers in the postwar period. In the late 1940s, the state considered workers buses were not public transit but designated group transport. Just after the war, the buses began operating without regulatory constraints, largely because state concession authorities failed to keep up with the constantly changing mobility needs in this turbulent period of rebuilding and reconstruction. When the transport situation finally improved, the Dutch government reinstated its interwar concession system. This raised barriers for industrial companies to charter or set up bus services for group transport. It also caused a lock-in for many: public transit operators had to deal with all the regulations. Influential employers with lobbying power managed to circumvent these by establishing semi-independent governing agencies to organize worker buses and align services with the Passenger Transport Act.[3]

The commuter bus system may have involved many different stakeholders, but employers were at the helm of this service. Hoogovens chartered local bus

3 Th. M. B. van Marle, *Het binnenlands vervoer na de bevrijding: 1944/45-1952* (The Hague: Nederlands Verkeersinstituut, 1953), 50-51.

companies and adjusted frequency, capacity, and routes to meet demand.[4] In contrast, Philips established the Industrial Personnel Transport Eindhoven Region (Vervoer Industrieel Personeel Rayon Eindhoven, VIPRE) in 1948, with stakeholders including state traffic inspectors, public transit companies, and private bus companies. These stakeholders cooperated closely with Philips to serve its mobility needs.[5] Limburg, with a longer history of miner buses, had two governing bodies. In response to the 1938 Passenger Transport Act, Limburg bus companies had established the Workers Transportation Union (Vereenigd Arbeidersvervoer, VAV), representing bus companies at meetings with the state traffic inspector. During the Second World War, another agency, the Miner Transportation Committee (Commissie Mijnwerkersvervoer), held monthly round-table meetings with the Workers Transportation Union, mining companies, and miner unions—together organizing daily buses for more than 10,000 workers.[6] Twente's textile industries and transport companies also collaborated in an overarching body. Starting in 1955, the Central Bureau for Industrial Personnel Transport Enschede (CBIPE) ensured that workers' bus commute did not stagnate, thus protecting the industry's interests.[7] In Rotterdam, shipping sector association (Scheepvaartvereniging Zuid, SVZ), tasked with dock workers' social welfare, organized buses subsidized by port companies.[8] The mining unions represented workers in a formal governance body. SVZ acted on behalf of dock workers in Rotterdam. Elsewhere, employers were in control: in Twente, employers and bus companies participated in the CBIPE without workers' representation as did Philips in the VIPRE. Only in the mining region were workers structurally represented in bus governance by the unions. In other cases, and in contrast to the 1920s and 1930s, employers rather than workers were the driving forces in organizing bus transport to serve their interests.

4 "Het komt in de bus," *Samen: Maandblad voor Directie en Personeel van Hoogovens en Nevenbedrijven* 26, no. 7-8 (1957): 128-129.

5 Philips Company Archives (PCA hereafter), Archive no. 642.5, Inv. no. 882, Correspondence 31 December 1948, 2-3, and Correspondence 20 June 1953, 1-2.

6 Regionaal Historisch Centrum Limburg (RHCL hereafter), Archive no. 17.05H, Oranje-Nassau Mijnen: Vervoer, 1903-1978, Inv. no. 46, Vervoer mijnwerkers per autobus, 1953-1972, newspaper clipping "170 bussen rijden dagelijks naar het Zuid-Limburgse kolenfront." *De Nieuwe Limburger*, 29 October 1955; *Wat moet u weten van het reizen met de VAV (Verenigd Arbeiders Vervoer) bussen* (Heerlen: Gezamenlijke Steenkolenmijnen in Limburg, 1953); "Busvervoer van mijnwerkers tot op de minuut geregeld," *Limburgsch Dagblad*, 16 December 1953, 5.

7 Stadsarchief Enschede (SE hereafter), Archive no. 61, Inv. no. 2, Verslag Personeelsvervoer 1955-1960, 1-2, 8-9.

8 Nijhof, *Gezien de dreigende onrust*, 138-143.

7.2 The Social Aspect of Bus Commuting

Companies used bus transport as an instrument to aid workers devoid of mobility options, as well as achieve an economic benefit. Later they also preferred commuting over migrating. This resembled how late nineteenth and early twentieth century employers and reformers considered railway expansion: reducing cost barriers to trains for working-class commuters enabled them to travel from their countryside homes to urban factories rather than forcing workers to move into packed, impoverished urban neighborhoods. As we have seen, interwar employers accepted workers' commutes as a necessity. After the war, however, industrial employers encouraged workers to commute because of housing shortages. Their policy to encourage commuting also fitted ideas about how to increase workers' productivity. Postwar employers wanted workers to continue living in the countryside without having to leave their family and community. The commuter bus catered to this (new) family idyll. Through the media outlets, mining operators tried to convey how upscaling transport served to "preserve Limburg against potential proletarianization." As with most forms of labor remuneration and company welfare, however, therein lurked a double motive. Managers aiming to preserve workers' so-called moral hygiene, presumed that workers increasingly preferred a healthy life with a little house and garden in the country. According to the director of State Mine Maurits, workers increasingly desired not to leave their familiar rural environment and live among the "mass proletariat" in industrial agglomerations. He claimed miners from villages like Blitterswijck in northern Limburg preferred to be at home in a familiar environment after a hard day's work, "not trade his goats and rabbits in the backyard for a small flat in the city." The journalist concluded that workers chose an hour and a half in the bus twice a day so that they could tend their garden and chickens or chat at the local pub.[9]

Indeed, workers' family lives impacted their mobility choices. Postwar sociological studies confirmed workers preferred their village and commute over migrating to industrial (urban) growth centers—a spacious green environment over a cramped high-rise apartment near work. Hoogovens thought commuters would move to near the factory once postwar housing shortages were resolved. When the steel company offered "its surplus houses" to workers

9 RHCL, Archive no. 17.05H, Inv. no. 46, 1953-1972, newspaper clipping "170 bussen rijden dagelijks naar het Zuid-Limburgse kolenfront." *De Nieuwe Limburger*, 29 October 1955; "Wat leert een bezoek aan de mijnstreek? De productie moet en kan omhoog," *Amsterdamsch Dagblad*, 7 July 1945, 3; Gerard Groot, "Massale volksverhuizing elke dag opnieuw in Nederland: Speciale busdiensten der bedrijven halen hun arbeiders van heinde en ver," *De Tijd*, 10 December 1955, 13.

in the 1950s, however, much to the surprise of its social services department, the commuters declined: "he would miss the wheeling and dealing in the village, the garden, and his wife wouldn't leave her relatives."[10] Workers from the north of Zeeland province commuting to Rotterdam responded "very emotionally" when asked about moving. Sociologists reached similar conclusions in 1957: Zeeland's workers would not leave their island Tholen and their community, because life was cheaper there, and they dreaded the unfamiliar anonymity of city life.[11] *Spil en Spoel* reported in 1962 that rural folk working in Enschede's textile factories were "attached to their own piece of land, even if it was far from work. They would never swap it for a house near the factory." The magazine observed "in an increasingly hectic world, the peace and quiet of the village weighed heavier than the downside of commuting."[12]

Buses bridged two social forces: employers who wanted cheap labor and workers who wanted to live in their communities. According to *Spil en Spoel*, "the bus connects the two worlds of the commuter," the factory work-floor and family life in the country, as Marten van der Ven wrote: "The minute he steps into spinning mill Oosterveld, he is a spinner. When he gets home, he is greeted by his family." And, "Overdinkel is the world where he lives and belongs; Enschede is the other world, where he goes to work every day. Commuting separates these two worlds. By choice, and if it was up to Mr Van der Ven, it would stay that way." For him, the magazine continued, rural Overdinkel had more to offer than urban Enschede: people knew him by his first name, the social life, "the freedom of a house with a garden," "the pigeon club," and local soccer club ASVO.[13] Lambooy and Mak, conducting a survey on the social impact of commuting at Hoogovens, observed that workers preferred commuting over migrating, not so much because they were "attached to their community." Rather they were "attached to living in a village" and "feared the consequences of living in an urban society." Lower rents, the opportunity to earn extra money on the land, leisurely pursuits like gardening, fishing, keeping animals, motivated life in the

10 J.G. Lambooy and J.H. Mak, "De pendelaar in woon-, reis-, en werkomgeving," (Koninklijke Nederlandsche Hoogovens en Staalfabrieken, n.d.), IV 1.

11 Zwemer, "De pendelarbeiders," 415.

12 "Zij komen van verre..," *Spil en Spoel: Maandblad voor het Textielvak* 13, no. 3 (1962): 1; "Voor- en Nadelen," *Spil en Spoel: Maandblad voor het Textielvak* 13, no. 3 (1962): 2. Whereas this image catered to the ideal of corporations and industrial managers—countryside lifestyle was considered beneficial for workers' health and thus productivity—this view was partly confirmed in sociological studies. Sociologists, however, also expressed a more cautionary tone, because long distance bus commutes came with adverse effects.

13 "Pendelen," *Spil en Spoel: Maandblad voor het Textielvak* 13, no. 3 (1962): 4-5.

country. They feared the conditions in apartment blocks, social isolation, higher rents, and limited scope for children.[14]

Not all workers had such room for negotiation. Migrant workers without unions to represent them had little choice. They were often deliberately housed in isolated company-owned barracks or hostels. Hoogovens preferred collective housing, to make workers more "accessible" to social workers and church organizations.[15] Industries built compounds like Casa Cortina in Almelo for Spanish, Italian, and later Turkish workers; Casa del Norte in Heerhugowaard (Hoogovens) for Spanish workers; and El Prado (Eindhoven) and El Pinar (Maarheeze) for Philips' migrant workers. According to migration historian Jaap Vogel, the barracks were also the materialization of a then widespread assumption by government and industry that they were temporary workers. Government policy was not aimed at integrating them into Dutch society. It expected they would eventually return home once their labor was no longer needed. Accommodation in isolated compounds was meant to maintain a certain order and discipline, and potentially accelerate social upliftment. Nonetheless, the isolated locations, coupled with limited mobility options, made these living quarters what Vogel tellingly described as "inhabited islands in an industrial landscape." Buses helped unlock them, I argue.[16]

7.3 Control Techniques and Strategies

Company buses comprised a socio-technical system that enabled companies to control and curtail workers' mobility. Philips managers claimed in 1953 that because VIPRE buses were "centrally governed and supervised," the number of absenteeism hours dropped to a minimum, from 1,800 in 1948, before the VIPRE buses, to 79 hours in 1952.[17] In other words, company buses may have helped attract cheap labor from afar, but also enabled companies like Philips to reduce latecomers and absenteeism.

Reliability was key for controlling the workforce—including workers' commute. In a 1957 interview, Van Eekhout of Hoogovens explained, "Everything

14 "De pendelaar," IV 3, VI 6-7, 9, Appendix 7, Appendix 9-10.

15 Melchior van Elteren, *Staal en Arbeid: Een sociaal-historische studie naar industriële accomodatieprocessen onder arbeiders bij het desbetreffend bedrijfsbeleid bij Hoogovens IJmuiden, 1924-1966* (Leiden: E.J. Brill, 1986), 783, 887.

16 Jaap Vogel, *Nabije vreemden: Een eeuw wonen en samenleven*, Cultuur en Migratie in Nederland (The Hague: SDU Uitgevers, 2005), 86-87, 89, 125, 127, 130-132.

17 PCA, Archive no. 642.5, Inv. no. 882, Correspondence 20 June 1953, 4-5.

is laid down in regulations. Each bus driver is given their route map and rules. And if everyone sticks to these, all goes well."[18] Hoogovens believed that "smooth transport" only thrived with "rules of the game" and compliance by bus drivers and "holders of bus cards" deemed employees eligible to use commuter buses.[19] Controlling time, cost, and workers' travel behavior was essential for this bus system to function.

The buses had to be up to standard and well maintained to ensure workers would use and could rely on them. Philips built a fully equipped parking garage for its VIPRE fleet, hired mechanics for daily repairs and maintenance, and installed on-site gas pumps.[20] Most industrial companies chartered their buses from local bus operators. The Miner Transportation Committee, CBIPE, and Hoogovens had binding contracts with private (charter) bus companies. To safeguard standards, these contracts stipulated that buses were clean, safe, and properly maintained, under penalty of contract termination.[21]

Workers frequently complained about the poor quality of bus materials, especially in the early development phase between the late 1940s and early 1950s. Complaints ranged from no heating in the winter months, to dirty interiors and poor maintenance. A Belgian woman said in 1949 that she found the bus commutes to Philips unpleasant, because of the poor state not only of the Belgian roads, but also the buses. "The cold creeps up your legs in winter," through the cracks in the floor and there was no proper heating. Textile workers complained about the poor state of commuter buses as late as the 1950s.[22] Although CBIPE successfully encouraged bus companies to improve their standards, charter bus companies Leguit, Zinger, Smithuis, and Snoeyink were notorious. Their buses frequently broke down, affecting the service schedule, and workers missed connecting buses, leading to time loss for factories. Passengers suffered from drafty buses and exhaust fumes, lack of heating in winter, dirty bus interiors, and faulty brakes. Workers protested to their superiors, who in turn notified CBIPE.[23]

18 "Het komt in de bus," 128-129.

19 Tata Steel Central Archives (TSCA hereafter), Archive no. 4338, Personeelsvervoer, Inventory no. RA-12871, newspaper clipping "Hoogovenexpres opgeheven: Bussen tussen station Beverwijk en ons bedrijf." *De Grijper*, October 1957.

20 PCA, Archive no. 728.14, Inv. no. 3; SE, Archive no. 61, Inv. no. 1, Contracten (1955).

21 TSCA, Archive no. 4349, Vervoer van en naar Beverwijk, Inv. no. RA-12877; TSCA, Archive no. 4344, Personeelsvervoer per autobus (HFD), Inv. no. RA-05831; "De pendelaar," 23.

22 PCA, Archive no. 642.5, Inv. no. 882, newspaper clipping "Dagelijks volksverhuizingen van België naar hier: Lange rijen autobussen voeren fabriekspersoneel aan," *Het Parool*, 26 October 1949.

23 SE, Archive no. 61, Inv. no. 5, Correspondentie Fabrikantenkring, 1955-1967, Tekortkomingen van vervoerders; Bedrijfsarchief, no. 17II, Correspondentie met beide commissies, 1955-1961, 15 January 1964, 1; Bedrijfsarchief, no. 6, Verslagen Vergadering Commissies 1955-1963, 2.

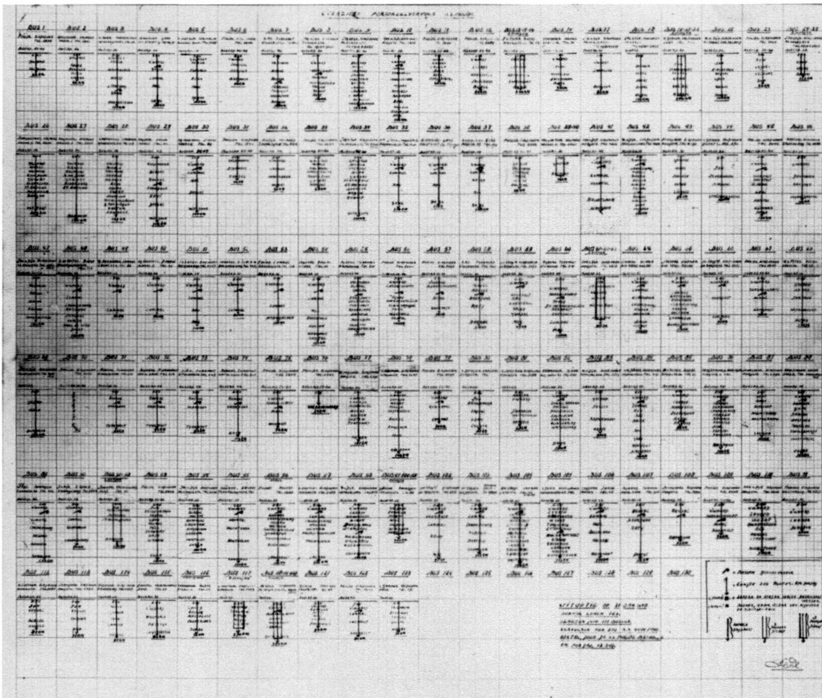

Figure 19: Draft of a detailed 1948 plan showing Philips VIPRE bus routes with assigned stops in villages around Eindhoven. (Source: Philips Company Archives)

The commuter bus system owed its success to detailed planning and alignment with industry and workers' needs. In 1953, *Limburgsch Dagblad* marveled at the scale of Southeast Limburg's bus system, driving about 10,000 miners to the mines every day, and praised its "wondrous accuracy." Commuter buses ran smoothly over the Limburg hills. Timetables and routes through the rural towns and hamlets were planned down to the minute and deviation was strictly prohibited.[24] Similarly, Hoogovens drivers were forbidden to stop at other places than in the timetable. The company bus thus prevented idleness and ensured no time was wasted during workers' commute.[25]

Via bus timetables and routes, employers gained control over workers' travel time in terms of duration and timing adjusted to the factory's rhythm of work shifts. Writing to the Ministry of Traffic and Public Works, VIPRE director P. Dekker emphasized the need for methodically planned commuter buses (fig. 19): "It is of utmost importance that people arrive at the factory in time. Because

24 "Busvervoer van mijnwerkers," 5.

25 TSCA, Archive no. 4338, Personeelsvervoer, Inventory no. RA-12871, newspaper clipping "Hoogovenexpres opgeheven: Bussen tussen station Beverwijk en ons bedrijf," *De Grijper*, October 1957.

most work is done at the conveyor belt, several people placed at such a conveyor belt coming in late, cause inestimable damage."[26] This led to the meticulous planning and adjustment of bus timetables to shifts and anticipated potential delays. A 1960s sociological report on commuters from rural North-Holland and West-Frysia working at Hoogovens described this careful timing. Buses had to arrive half an hour before a shift starts and half an hour before departure at the end of a shift. This had a dual purpose: workers could shower and/or change clothes—it was not permitted to board buses in grubby work wear—and gave some slack in case of delays or bad weather. On-site buses brought workers to the main gate, where company police conducted the usual inspection. Once they passed inspection, workers waited at the bus stop, "usually in orderly queues."[27] This time-discipline was to eliminate the risk of workers wasting time, ensuring maximum efficiency and unhampered movement.[28]

Strict regulations on how to travel by bus were inculcated into workers' minds. Passengers in company-organized buses—like employees on bicycles and mopeds—had to comply with regulations set by their employer. Violating the regulations risked suspension from using these buses (thus forcing workers to commute by more expensive public transit or revert to cycling) and disciplinary measures. These regulatory efforts aimed to model employees into docile bodies who would not hamper the minutest bus operations. To control crowds of mineworkers getting on and off buses and avoid stagnation at bus stops in Limburg, the VAV union booklet instructed miners to board the first available bus at their stop, only at designated bus stops, and on the right side of the road. As drivers were commonly out in the early morning or late at night—and given that roads at the time were poorly lit—they could not oversee all sides of the bus and thus ran over jaywalking miners. Stepping off the bus had to be done in an orderly fashion—not only to prevent hold-ups, but also for workers' and others' safety, the managers believed. Miners could only use designated entrances and exits, and certainly not jump off moving buses, like in the days of improvised buses.[29]

26 PCA, Archive no. 642.5, Inv no. 882, Correspondence 20 June 1953, 4-5.

27 "De pendelaar," 24.

28 Foucault argued that the "time-table," originally a monastic model, spread into schools, factories, and hospitals to serve three purposes: "establish rhythms, impose particular occupations, regulate the cycles of repetition." Based on the principle of non-idleness: "[It] was forbidden to waste time ...; the time-table was to eliminate the danger of wasting it—a moral offence and economic dishonesty. In the correct use of the body, which makes possible a correct use of time, nothing must remain idle or useless." However, "Discipline ... arranges a positive economy ... maximum efficiency." Michel Foucault, *Discipline and Punish: The Birth of the Prison* (London: Penguin/Random House, 2020 [1977]), 139, 143-144.

29 RHCL, Archive no. 17.05H, Inv. no. 46, Minutes 24 September 1946; *Reizen met de VAV*, 7.

A company photo (fig. 20) taken in 1955 at Hoogovens shows commuter buses parked just outside the factory entrance. The blast furnace site with coke furnace chimneys is in the background and the guarded gate is where workers had to show their identify card before entering the site. The bus shelters in the lower left corner offered workers protection against rain and wind and possibly shade in the summer. What stands out are the regimented queues of buses and workers, lining up without pushing, bus card in hand ready to show the bus driver. Although not always so orderly, this is what employers aimed for with their travel regulations—and sought to capture in their company photography. Hoogovens also presented norms about proper travel behavior in the form of catchy, easy to remember rhyming slogans. A booklet *What bus or train do employees take to Hoogovens?* (1951) was handed out to commuting employees, telling them to behave like a "gentleman in traffic," just like their cycling counterparts. On the very first page was a short poem describing gentlemanly behavior:

"A man who was born a real gent,
Would never go to work by train
That is, until one day he noticed
A Hoogovens workers train, and then spoke:
Oh verily, will you all listen
Indeed, nowhere else did I see
A tangle of people succeeding
In boarding a train without fuss
Then seating themselves in their seats
And doing all this without spitting.
A roughneck not nearly a gent,
The terror of ev'ry bus driver,
Pushed everyone rudely aside,
And shouted: 'Me first and then you!'
Before throwing himself on the seat.
Then one day traveling the Hoogovens bus
A standing commuter addressed him
And offered his coat for comfort.
The scoundrel was finally dumbfounded
And was seen to behave ever since."[30]

30 TSCA, Archive no. 4337, RA-04155, "Met welke bus of trein gaan werknemers naar de Hoogovens?," 20 May 1951, ("Een man, die als heer was geboren,/ Die ging uit principe nooit sporen./ Tot hij op een dag/ Een HO-treintje zag./ Hij sprak toen tot elk die 't kon horen:/ 'k Zag nooit in

Figure 20: Hoogovens employees lining up to board commuter buses at the end of their shift. IJmuiden, 1955 (Source: Tata Steel Central Archives)

The booklet included many such messages, sprinkled between timetables and routes. Hoogovens made it clear that employees were expected to behave properly on the bus, keep it clean, and not leave litter: "Buses and trains are clean and cleaning costs time and money. As a passenger, make it your motto 'In the bus and in the train, I want to behave like a gentleman.'"[31] Similarly,

m'n leven zo'n kluwen/ Van mensen, die zonder te duwen/ De treinen bestegen,/ Op hun plaatsen neerzegen/ En op de reis niet in 't rond gingen spuwen./ Een ruwaard, haast niet om te sussen,/ De schrik van 's lands autobussen,/ Drong ieder opzij, Sprak: 'Ik eerst dan jij!'/ En smeet zich dan breeduit op het kussen,/ Eens zat hij met stomheid geslagen,/ Heeft zich nadien ook wat netter gedragen/ Toen in een HO-bus/ Een heer aan de lus/ Zei: 'Kan mijn jas U als deken behagen?")

31 TSCA, Archive no. 4337, RA-04155, "Met welke bus of trein gaan werknemers naar de Hoogovens?" 20 May 1951, 2-19.

the VAV booklet advised Limburg's miners to avoid "unpleasantness." Miners were responsible for getting to the designated bus stops in time to ensure a seat and then getting to the mine shaft on time. The booklet reminded miners that the driver supervised whatever went on in and around the bus, and that they risked exclusion from the bus service if they did not comply. Miners must not insult the drivers, obstruct emergency exits with their duffle bag ("pungel"), or spit on the floor.[32] Promoting self-discipline was how employers sought to ensure unhampered bus transport.

Pricings schemes were also an important tool. Company bus pricing schemes indicated the fares workers paid within a certain radius from sites. According to the VAV booklet, the zoning stabilized otherwise variable bus fares for miners, and for the industry, prevented "unmotivated growth of bus transport on short distances." The further away workers lived, the more financial compensation they received for using the commuter bus—up to 80 percent. Whether miners qualified for compensation was determined by calculating the shortest commuting distance by bicycle, taking into consideration the terrain traversed. For cycling distances under 4 km, workers paid 3.85 guilders for a weekly bus card; workers living at 4 to 7 km cycling distance over flat terrain, paid 2.90 a week; miners cycling routes over hilly terrain or distances beyond 7 km paid less: 1.90 a week. Young workers under twenty with an hourly salary lower than 64 cents: they only paid 95 cents a week for distances over 7 km.[33]

Allowances differed by company. VIPRE emphasized that Philips free commuter buses were a travel allowance in-kind, not a right for every employee. They were only for workers living outside a 12-km radius, in designated areas where Philips wanted to attract workers otherwise not able to reach the factories in Eindhoven. Miners, whatever the travel distance, paid a modest share of the VAV fares. Director Dekker explained that, wherever public transit alternatives fell short—i.e. not adjusted to workers' shifts or residential locations—VIPRE brought solace. To keep costs down, however, they reduced the number of stops within a certain radius of Philips sites, prohibiting workers living in Eindhoven and surrounding villages from using commuter buses. VIPRE assumed workers within the 12 km radius had sufficient access to mobility alternatives like walking, cycling, or public transit.[34] Hoogovens had a similar policy. According to its magazine *De*

32 *Reizen met de VAV*, 7.

33 *Reizen met de VAV*, 1-5; "Busvervoer van mijnwerkers," 5.

34 PCA, Archive no. 642.5, Inv. no. 882, Correspondence 20 June 1953, 3-4, Correspondence 27 November 1970, 2-3, Correspondence 14 April 1971, 1-2, Correspondence 27 November 1970, 2.

Grijper: "it is not possible to let everyone—for example contractors' staff or workers from Beverwijk who want to leave their bicycles at home when it rains—make use of the buses without hampering services." The new service could "only be used by employees who used to travel by train and live 'outside the vicinity' of the company" like in Beverwijk, Heemskerk, Velsen-Noord, IJmuiden, Driehuis, and Santpoort-Dorp.[35] Hoogovens steel workers in the "travel area" could either get a bus card and timetable or request a personalized bus card.

Timetables and routes were another instrument. Industrial employers often plotted tables and maps based on the location of workers' homes, distinguishing workers within and below the "sufficient accessibility" threshold (fig. 21). To reduce costs, commuter buses were only for workers within the domain of insufficient accessibility, lifting them above the threshold. Others were excluded because employers believed they could pay for their commutes themselves. Workers considered having insufficient mobility options saw their mobility enhanced through company intervention, while those with allegedly sufficient options had to fend for themselves.[36]

Bus drivers or controllers served as gatekeepers. Workers had to show the driver or regulator a centrally distributed bus card when boarding. Their bus card doubled as identification card, proving their eligibility to use commuter buses. Valid for a week, the personalized bus cards showed workers' name and number, and sometimes the color of the card indicated their shift. Leaflets and bulletin boards frequently reminded miners to collect their personal bus card before the start of a new working week. Without a VAV bus card, miners would have to pay the full fare or find other means of travel.[37] Hoogovens bus cards had a photo, to prevent workers sharing cards with co-workers.[38] The same rules applied to textile workers using CBIPE buses. Because different cards were confusing, the CBIPE in 1956 centralized the distribution with the employer, changed the bus card format, and shifted to bus cards valid for a year—only for workers in affiliated factories and living in designated areas. CBIPE director Slegt urged bus drivers to thoroughly

35 TSCA, Archive no. 4338, Inventory no. RA-12871, newspaper clipping "Hoogovenexpres opgeheven: Bussen tussen station Beverwijk en ons bedrijf." *De Grijper*, October 1957.

36 Ibidem. For similar ideas, see: Karel Martens, *Transport Justice: Designing Fair Transportation Systems* (New York/London: Routledge, 2016); Sheller, *Mobility Justice,: The Politics of Movement in an Age of Extremes* (London/New York: Verso, 2018), 27.

37 RHCL, Archive no. 17.05H, Inv. no. 46.69, Minutes 24 September 1946; *Reizen met de VAV*, 1-5; "Busvervoer van mijnwerkers," 5.

38 TSCA, Archive no. 4338, Inventory no. RA-12871, "Hoogovenexpres opgeheven"; "Het komt in de bus," 128-129.

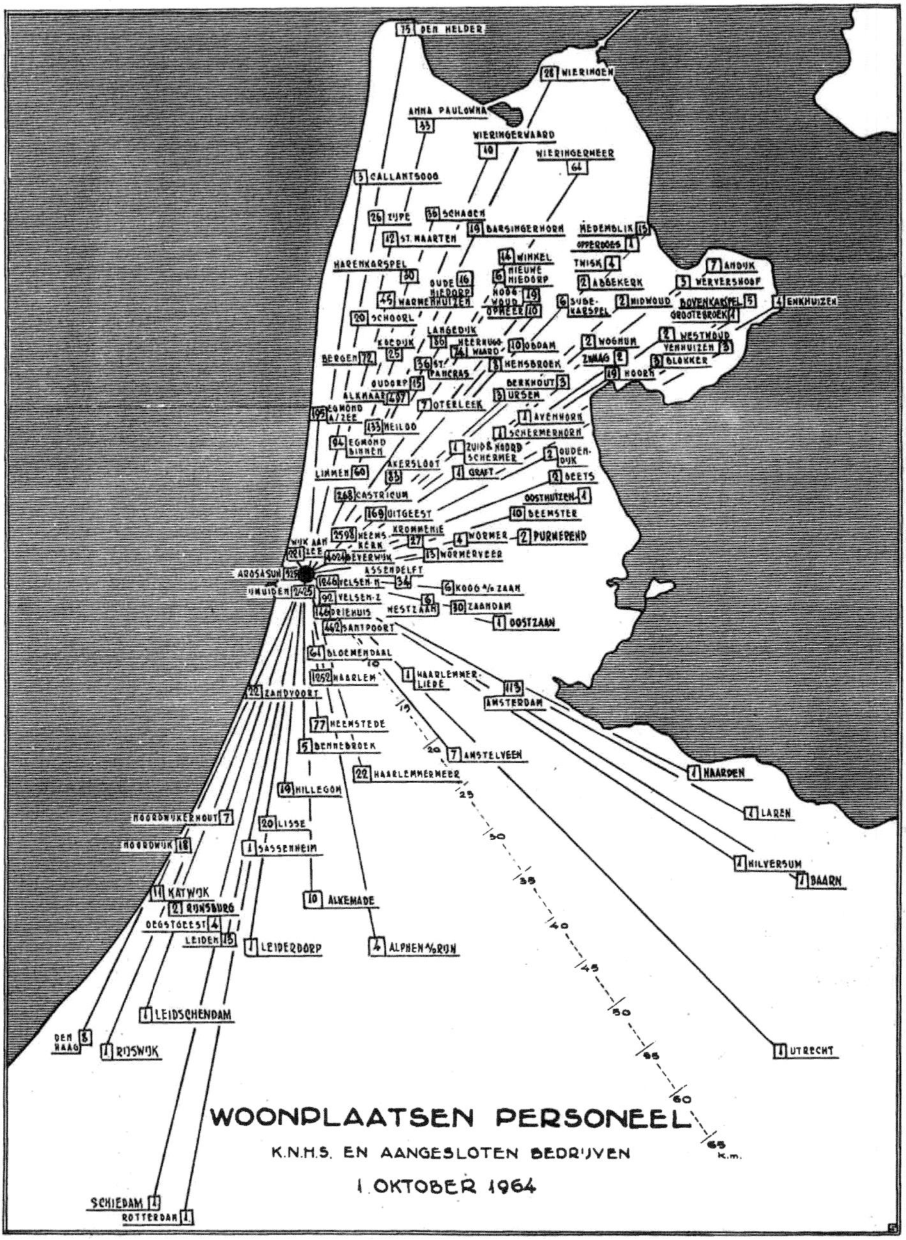

Figure 21: Postwar map showing the geographic spread of Hoogovens employees who commuted by bus. 1964 (Source: Tata Steel Central Archives)

check the personal details, the color of the card indicating shift time, and the written adjustments in pencil on the card.[39] A Hoogovens reporter on the commuter bus saw that company controllers frequently halted the buses to check passengers' cards. "We had an inspection, but there were no fare dodgers, and everyone had their bus cards." CBIPE inspections often found passengers who had allegedly forgotten their cards or, because they worked in non-affiliated factories, were trying to dodge the higher public transit fares.[40] These and other control mechanisms ultimately ensured companies an (cost-)efficient operation of commuter buses.

Finally, employers segregated workers according to ethnicity and gender. Labor historians have shown this was common practice on the shopfloor with men and women working in separate spaces, often motivated by a patriarchal desire for social control over women's sexuality, for example.[41] This also applied to buses. A 1959 government report noted that "to prevent fighting for the best seats and unpleasantness," Moluccan workers were assigned seating in company buses. "They were given seats—as they requested—next to each other, therefore not encouraging contact with Dutch people." Many companies ran separate buses for Dutch and Moluccan workers, partly due to the different routes these buses had to take to the diverse housing locations, but partly to avoid any hassle.[42] Philips even exploited separate buses for men and women workers. In 1953, VIPRE reported that men and women traveling in the same bus had caused "unpleasantness." Director Dekker did not elaborate on what had occurred or what actual problem was now solved by segregating men and women.[43] Workers had no choice but to comply. Special 'managers' made explicitly clear that all passengers had to comply with the rules on each VIPRE bus. These controllers were

39 SE, Archive no. 61, Inv. no. 6, Verslagen vergaderingen commissies, 1955-1963, "Nieuw plan vervoersorganisatie" (1956), 2, Inv. no. 8, Controle-rapporten weekkaarten, 1957-1958, Correspondence 22 August 1956, p.1; Correspondence 20 February1957, 1; Correspondence 19 March 1957, 1; Correspondence 20 March 1957, 1; and Correspondence 3 April 1957, 1-2.

40 SE, Archive no. 61, Inv. no. 6, Verslagen vergaderingen commissies, 1955-1963, "De organisatie van het arbeidersvervoer en de taak van het Centraal Bureau" (1955), 4-5, Verslagen vergaderingen commissies, 1955-1963, "Nieuw plan vervoersorganisatie" (1956), 2, Inv. no. 8, Controle-rapporten weekkaarten, 1957-1958, Correspondence 22 August 1956, 1; Correspondence 20 February 1957, 1; Correspondence 19 March 1957, 1; Correspondence 20 March 1957, 1; and Correspondence 3 April 1957, 1-2; "Het komt in de bus," 129.

41 Gertjan de Groot, *Fabricage van verschillen: Mannenwerk, vrouwenwerk in de Nederlandse industrie (1850-1940)* (Amsterdam: Aksant, 2001), 28-29, 33-34, 103.

42 *Ambonezen in Nederland: Rapport van de Commissie ingesteld bij besluit van de Minister van Maatschappelijk Werk*, (The Hague: Staatsuitgeverij, 1959), 58-59.

43 PCA, Archive no. 642.5, Inv. no. 882, Correspondence 20 June 1953, 4-5, "Rapport: Oprichting en werkzaamheden van de N.V. V.I.P.R.E." (1953), 4.

co-workers who received 5 guilders a week for this task. Bus drivers were all men, whereas controllers were men in the men's buses, and women in women's buses. These controllers ensured proper behavior. While all (Belgian) women sang, then waved and shouted as they passed a bus with young men going to Philips, "the bus manager, the hard-working woman who keeps everything in order, and is a bit like a mother, put a stop to that."[44] Often bus drivers fulfilled a similar role as gatekeeper. It is worth noting such ethnic and gender segregation did not exist in public buses.

For employers, company buses had several advantages over public transit services. Unlike on public buses, company bus capacity and frequency were adjusted to the number of employees using these buses, to keep costs low and services running smoothly. A strict order of people who could or could not board was thus vital. It also reified a social stratification in mobility access based on people's labor relations with the company and their geographic location. Policies determined this stratification, imposed by controllers and bus drivers. The most important task for controllers was to inspect all bus cards and keep out unwanted passengers. A *Philips Koerier* reporter, writing about the buses to Belgium in 1946, even feared he could not board a VIPRE bus, because "strangers were strictly forbidden."[45] Employers drew strict boundaries regarding usage. This mono-functional use of commuter buses enabled but also relied on strict regulative governance of who was (not) allowed in them—explicitly distinguishing employees from non-employees and dividing them according to function, geographic location, and gender.

With this system, industrial companies constructed and empowered specific mobile subjects, primarily motivated by utilitarian, cost-benefit considerations. With postwar commuter buses, employers prioritized a particular kind of employed, able-bodied worker with particular (gendered) skills from specific geographic locations with cheap labor. The tools to do so ranged from regulations booklets to pricing schemes and personalized bus cards—unlike public transit, where passengers could board as long as they had a valid ticket. The booklets and tickets were a very tangible materialization of companies' aims to control their workers' movements.

44 PCA, Archive no. 642.5, Inv. no. 882, Correspondence 20 June 1953, 4-5, "Rapport: Oprichting en werkzaamheden," 4, and Correspondence 27 November 1970, 2; "Susanneke werkt in Nederland," *De Vrije Philips Koerier* 3, no. 7 (1946) 2; "Het Personeelvervoer," *De Vrije Philips Koerier* 3, no. 8 (1946): 2; "In de bus," *Philips Koerier* 3, no. 44 (1946): 2.

45 "Susanneke werkt in Nederland," 2; "Het Personeelvervoer," 2; "In de bus," 2.

Conclusion

Postwar company buses represented increased control over worker's mobility not only compared to cycling, but also to earlier bus services. Employers appropriated governance over buses, taking over the initiatives by workers who had partly led in self-organizing commuter buses in the interwar period. In the postwar era, unlike the interwar period, workers had little say over buses. Employers bore the cost of travel, enabling low-income rural and migrant workers to gain employment further afield, but also entailing transport was on companies' rather than on workers' terms. And compared to urban workers, rural and migrant workers' off-site movements came under a higher level of company control. Bus riders were subjected to more top-down forms of control compared to cyclists and moped riders: companies considered the buses an extension of their territory. The meticulously planned company buses represented the materialization of company ambitions to control rural and migrant workers' movements.

Over time, buses became a greater part of workers' mobility, forcing workers' movements into a system of time-and-physical factory discipline. Although buses also offered workers a fast, flexible, reliable, and safe mobility alternative to public transit, employers could regulate workers' movements and align travel behavior with company interests and the rhythm of factory work better than with bicycles and mopeds.

Time-discipline played a pivotal role in the organization of company buses. E.P. Thompson's classic study, "Time, Work-Discipline, and Industrial Capitalism" (1967) shows that the rhythm of factory work became the guiding principle for how workers lived and worked when European and North American societies industrialized. This time management was crucial for labor discipline and industrial growth.[46] Recently, Marcel van der Linden summarized the time-discipline as concerning "the beginning, duration, and end of the working day, the timing of breaks, holidays, the numbers of activities that have to be performed during a certain labour time," and at the core of scientific management.[47] "Mobilizing Rural and Immigrant Workers" has shown this form of governmentality also applied to postwar commuter buses.

The 1970s, however, marked a turning point, a transition away from this paternalistic model and the postwar welfare state towards a more neoliberal view of workers' mobility, as we will see in the next chapter.

46 Edward Palmer Thompson, "Time, Work-Discipline, and Industrial Capitalism," *Past & Present*, no. 38 (1967): 56-97.

47 Linden, "Work Incentives and Forms of Supervision," 482-483.

8 Leaving Workers to their Own Devices during Deindustrialization, 1970-1990

When in 1975 the 22-year-old jobseeker K. appeared in court for declining a job that he considered poorly accessible, the Central Board of Appeal found his reasoning ungrounded. The board ruled: K. had to accept the job and find a way to get there or lose his unemployment benefits. Anyone who like K. chose to reside in a place with limited job opportunities bore sole responsibility for "taking necessary measures" to get to potential places of work.[1] The case became a legal landmark case that established the new norm. The case ruled that workers were now solely responsible for getting to their workplace. The ruling was a bellwether for the neoliberal economic order.

The shift took place at a time of recession and global restructuring. Amid a global wave of deindustrialization, heavy industries largely disappeared or declined. Closures, outsourcing, and automatization resulted in mass-scale lay-offs in the Global North—and in the Netherlands as well. Shipyards, textile factories, and heavy industry faced bankruptcy and were forced to lay off hundreds to thousands of blue-collar workers. Entire industries collapsed. By the 1960s, Twente's textile industries left 30,000 men and women jobless, and in 1974, Limburg's last mine closed with a loss of 75,000 direct and indirect jobs. Hoogovens and Philips would follow later. The postwar era of industrial growth and high employment belonged to the past, as unemployment rates soared for the first time since the Second World War. By the 1982 elections, symbolizing a right turn in Dutch politics, the specter of one million unemployed loomed. Many breadwinners who had worked in heavy industries for decades were left without jobs. Unemployment also hit women and young people like K. Mobility barriers exacerbated these people's gap to the labor market. The industrial collapse in the period of deindustrialization has been well described. Thus far, it has not been understood in terms of workers' loss of mobility.

1 Nationaal Archief (NA hereafter), Archive no. 2.15.62, Directoraat-Generaal voor de Arbeidsvoorziening van het Ministerie van Sociale Zaken en Werkgelegenheid, 1970-1990, Inv. no. 2199, "Handleiding Passende Arbeid" (1977), C.R.v.B., 27 March 1975, R.S.V. 1975 no. 341.

Bek, Patrick, *No Bicycle, No Bus, No Job: The Making of Workers' Mobility in the Netherlands, 1920-1990*. Taylor & Francis Group 2022
DOI: 10.5117/9789463723183_CH08

Around the time of K.'s appeal, the disadvantage of being car-less in a car-centered society became visible. For the Dutch case, this still has to be analyzed. For the U.K., sociologist Colin Pooley has argued in 2016 that, as British society became more car dependent, those without a car increasingly experienced barriers to get around, exacerbated by the erosion of public transit and heightened expectations of high mobility in modern society.[2] Despite growing car usage, nearly half of British households was still without a car by 1970, as urban historian Simon Gunn has shown.[3] Around 1970, these problems were also signaled by critical economists and geographers in the U.K., U.S., and Sweden. They began to understand—and coin—the new phenomenon as "transport poverty," extending their scope for the first time to social groups other than the working population. Not having a car in modern society had become synonymous with being 'transport poor,' as cycling was not a viable option for everyone and eroding public transit failed as an affordable alternative. I will show—as scholars have thus far not addressed—that observations by sociologist Enne de Boer in 1976 led to a debate in the Netherlands on forced car usage and what De Boer coined the "accessibility crisis for the car-less" in the following years.[4]

"Leaving Workers to their Own Devices" shows that when immobility increased, forced car ownership as a vehicle to access work became normalized—even in the Netherlands. It rendered the bicycle important for shorter commuting distances. For longer distances, public and company support for workers commutes was no longer self-evident.[5] For the first time, scholars began to conceptualize these problems, though with little impact on government and company policies. It was a fundamental shift. For one, neither the state nor employers felt the same need and/or responsibility for workers' commute as before. They now deemed it the workers' responsibility. Consequently, jobseekers like K. had to fend for themselves. K's case marks—and this chapter traces—this shift.

2 Colin Pooley, "Mobility, Transport and Social Inclusion: Lessons from History," *Social Inclusion* 4, no. 3 (2016): 100-109.

3 Simon Gunn, "People and the Car: The Expansion of Automobility in Urban Britain, c.1955–70," *Social History* 38, no. 2 (2013): 236-237.

4 Enne de Boer, "Mobiel en niet-mobiel: Een verkenning van de sociale betekenis van ons vervoer," in *Stedelijk verkeer en vervoer langs nieuwe banen? Preadviezen voor het symposium, 5 oktober 1976*, ed. J. Overeem (The Hague: Stichting Toekomstbeeld der Techniek, 1976), 47-78.

5 De Boer, "Mobiel en niet-mobiel," 47-78; "Vervoersongelijkheid, relevantie en onderzoekbaarheid: Een verkenning van de internationale literatuur," (The Hague: Projectbureau Integrale Verkeers en Vervoersstudies, Ministerie van Verkeer en Waterstaat, 1980); *Transport Sociology: Social Aspects of Transport Planning* (Oxford: Pergamon, 1986).

8.1 Employers Withdraw

Scholars have described the 1970s process of deindustrialization. It caused not only job losses, but a wider social disintegration of the societies and economies built around industrial bases. According to American labor historians Jefferson Cowie and Joseph Heathcott in the edited volume *Beyond the Ruins: The Meanings of Deindustrialization*, the closure of docks, factories, and mines in America challenged people's sense of community at the deepest level—a phenomenon also happening in regions with strong industrial bases in the Netherlands, like Twente's textile region and Limburg's mining region. Affected communities struggled to survive as the social structure in which they had thrived disappeared. It felt "like the end of a historical epoch." For a generation of millions of working men and women, what they may have experienced "as solid, dependable, decently waged work," in a long-term perspective, "really only last for a brief moment in the history of capitalism."[6] The wide-ranging company welfare programs typical of twentieth century industrialization now withered.

This also applied to company transport. Among the crumbling industrial infrastructures were company buses. In 1990, the Federation of Dutch Trade Unions (Federatie Nederlandse Vakbonden, FNV) reflected on companies' lack of support for workers' mobility. Before the 1970s, companies organized workers' transportation to make up for the labor market shortage and reduce cost barriers, since "workers' wages were too low to afford their own transport."[7] Some companies kept buses running, sometimes with downsized fleets. By 1990, twenty percent of almost 28,000 Hoogovens employees still used free company buses. Philips maintained shuttle buses connecting main transit hubs with Eindhoven factories and long-distance buses as far as Aken in Germany (100 km), Brussels (130 km), and Enschede (180 km). Shipyard companies like Wilton Fijenoord also had smaller buses for migrant workers in 1971. But in general, company buses played a less significant role in workers' mobility in the 1970s, even more so in regions where industrial

6 Jefferson Cowie and Joseph Heathcott, "The Meanings of Deindustrialization," in *Beyond the Ruins: The Meanings of Deindustrialization*, ed. Jefferson Cowie and Joseph Heathcott (Ithaca, NY: ILR Press, 2003), 1-4; Oliver Dinius and Angela Vergara, *Company Towns in the Americas: Landscape, Power, and Working-Class Communities* (Athens, GA: University of Georgia Press, 2011), 16-17. Dinius and Vergara note that since the 1970s, company managers were less willing to invest in company housing and social amenities.

7 Fons Tuinstra, *Woon-werkverkeer in schonere banen* (Amsterdam: FNV Centrum Ondernemingsraden, 1990), 33-35.

bases completely disappeared.[8] When textile factories closed, the CBIPE downsized its fleet and eventually stopped its operations in 1969.[9] When the mines closed in Limburg, transport companies also dissolved. In neighboring countries too, employers no longer subsidized mass-scale bus transport.[10] The fewer jobs available were now also less accessible.

This was even true—a point that needs emphasizing—for new and thriving industries. In newly developing industrial areas, companies failed to organize a commuter bus system. The Chamber of Commerce and Factories and the Transport Advice Group (TRAG) observed in Rotterdam's new Botlek-Europoort area that organizing workers' bus transport was problematic. Chemical and transshipment industries here offered 20,000 jobs to the regions' unskilled, semi-skilled, and skilled workers. Some companies did charter buses to attract workers from Rotterdam (40 km), Vlaardingen (30 km), and further away Bergen op Zoom (90 km), but often limited their services to a maximum of nine passengers or ran taxis. Others left employees to their own devices, occasionally providing commuting expenses. Companies did not manage, however, to come up with a cost-efficient, structural solution.[11] This failure represented a wider trend of declining company involvement in workers' mobility.

Workers did object to the shift. Their unions sought to protect workers' access to commuting options. In anticipation of the 1990 collective labor agreement, the union FNV urged Dutch employers to get involved more directly in workers' mobility by offering alternatives to cars—mostly to curb car usage. In *Woon-werkverkeer in schonere banen* (1990), FNV information

8 Philips Company Archives, Archive no. 728.14, Inv. No. 3, Philips Snelbusdiensten (April 1970), Routes Sneldiensten 7 June 1974, "Neem rustig de bus," newspaper clipping "Snelbusdienst uitgebreid: Voortaan vier lijnen," *Philips Koerier*, 10 January 1974; Tuinstra, *Woon-werkverkeer in schonere banen*.

9 Stadsarchief Enschede, Archive no. 61, Inv. no. 15, Correspondence 12 February 1969, and "Rapport uitgebracht Aan het bestuur van het Centraal Bureau Industrieel Personenvervoer Enschede, te Enschede: Jaarrekening 1968," 2.

10 Sophie Bouwens, *Over de streep: Grensarbeid vanuit Zuid-Limburg naar Duitsland, 1959-2001*, vol. 71, Maaslandse Monografieën (Hilversum: Verloren, 2011), 71, 119, 146.

11 NA, Archive no. 3.17.17.04, Inv. no. 1734, Stukken betreffende de bemoeienis met het onderzoek naar de verbetering van het woon-werkverkeer van en naar het Europoort- en Botlekgebied door middel van groepsvervoer, 1967-1968: Transport Advies Groep, 'Rapport', 1-4, 7-10, 20, 35. TRAG was commissioned by different stakeholders in the Europoort-Botlek industry, namely the Stichting Europoort-Botlek Belangen, De Metaalbond-afdeling Rotterdam, Chamber of Commerce and Factories of Rotterdam, and Stichting "Eilanden-gemeenschap Voorne-Putten en Rozenburg." TRAG collected data from 44 companies in the period 1967-1968, ranging from smaller to larger industries, from automated to labor intensive, from chemical to container terminals and metal industries.

officer Fons Tuinstra queried: whether employers' businesses were located near a train station; safely accessible by bicycle at night; and if working times were adjusted to public transit timetables. These were "questions that an employer did not usually need to answer," but as he pointed out, these were still problems "for the individual employee to solve." Tuinstra assumed that employers had not been concerned about these issues in recent decades, because due to increasing household incomes, housing and mobility had become "more of a concern for the people.[12]

The state also became aware of employers withdrawing. Like FNV, the government's advisory body, the Social Economic Council (SER) noted in its Handbook on Commuting (*Handboek woon-werkverkeer* 1991) how Dutch companies' direct involvement in workers' mobility declined in the mid-1970s. Employers' attitude of direct (paternalistic) involvement in workers' mobility shifted towards a more discretionary, indirect involvement via travel allowances. SER pointed to the "individualization of workers' transport." Rising wages and the desire to drive to work, allegedly invalidated the main argument for companies to provide company transport. Besides, for most companies, the cost per employee for company buses was significantly higher than a travel allowance based on public transit fares. This was why the government's advisory agency SER claimed companies had shifted from direct company support (via buses) to travel allowances.[13] Company support for workers' mobility thus appeared sensitive to companies' planning horizon and global economic fluctuations. In addition to mass lay-offs and relocation of production processes to low-wage countries in the Global South, companies seeking to cut costs also divested in workers' commute. The private car was promoted to fill this void.

8.2 Forcing Car Commuting as the New Normal

Historians have well documented the car industry's push for mass-scale automobility. Key in the messaging was how everyone could afford a car in the very near future. The campaign also reached personnel magazines. A 1965 cover (fig. 22) of textile industry magazine *Spil en Spoel* shows a car parked in the background, and an employee in the shadows gazing over the new car park. The visual message was a worker aspiring to owning a car one day: he

12 Tuinstra, *Woon-werkverkeer in schonere banen*, 7-9, 33-35.

13 Roel Meilof, *Bijlage: Wegwijzer woon-werkverkeer*, ed. COB/SER (The Hague: Commissie Ontwikkeling Bedrijven van de Sociaal-Economische Raad, 1991), 13.

(the breadwinner) could travel to work in comfort. Cars no longer belonged to "a certain class," editors claimed: "There are more and more employees' cars in the textile companies' carpark," exclaiming, "What was a fantasy 10 or 20 years ago, is now finally a reality: The Netherlands is under the car's spell!"[14] The American way of life was bound to reach the Netherlands as well.

Automobility, advertised by car boosters as a source of freedom, extending middle-class people's horizons in the era of affluence, linked the private car to value dimensions of modernity: social mobility, autonomy, and individuality.[15] U.S. car boosters generally claimed that private cars and highways emerged as the result of changing consumer preferences and autonomous individuals eagerly exercising their right to freedom of movement. So did the Dutch car lobby, which found unlikely bedfellows.[16] In 1977, journalists Hugo Arlman and Gerard Mulder discussed socialist filmmaker Jan Vrijman's propaganda movie "The Road is for EVERYONE" ("De weg is voor IEDEREEN") in the left-oriented weekly *Vrij Nederland*. The movie, commissioned by the Dutch Roadbuilders Association (NVWB), presented the car and its accompanying infrastructures as emancipatory vehicles for the "proletarian masses." Contrasting black and white film footage of interwar poverty and penniless benefit recipients with more recent shots of enormous parking lots near supermarkets and families enjoying outdoor recreation by car, Vrijman celebrated automobility. Car boosters like Foundation Road (Stichting Weg) and NVWB considered themselves "trailblazers of democracy," standing up for the interests of the "the little guy" and the "masses of workers," Arlman and Mulder wrote.[17] This trope of

14 "Na 20 jaar," *Spil en Spoel: Maandblad voor het Textielvak* 15, no. 10 (1965): cover, 2; "Een eigen auto ... Waar doen ze het van?," *Spil en Spoel: Maandblad voor het Textielvak* 15, no. 10 (1965): 10-11.

15 Stephan Rammler, "The Wahlverwandtschaft of Modernity and Mobility," in *Tracing Mobilities: Towards a Cosmopolitan Perspective*, ed. Weert Canzler, Vincent Kaufmann, and Sven Kesselring (Hampshire/ Burlington: Ashgate, 2008), 68.

16 Alan Walks, "Driving Cities: Automobility, Neoliberalism, and Urban Transformation," in *The Urban Political Economy and Ecology of Automobility: Driving Cities, Driving Inequality, Driving Politics*, ed. Alan Walks, Studies in Urbanism and the City (London/New York: Routledge, 2015): 4-20, here 11-14. In the Netherlands, Stichting Weg lobbied on the highest political level for more car space. The Stichting was founded in 1965 by the country's most important stakeholders in automobility: ANWB, BOVAG, RAI, Shell, ESSO, Chevron Petroleum Maatschappij, Nederlandse Vereniging van Wegenbouwers, Vereniging voor Bitumeuze Werken, Nederlandse Maatschappij voor Nijverheid en Handel. It also received support by semi-public institutions like Rijkswaterstaat and Netherlands' Chambers of Commerce, several national newspapers, and the car-owning middle class and small business owners.

17 Hugo Arlman and Gerard Mulder, "Een warmkloppend hart voor de noden van de massa's," *Vrij Nederland*, no. 44 (1977): 4-5; "Voor de mens als 'tool-using animal' is het gebruik van de auto heel natuurlijk," *Vrij Nederland*, no. 44 (1977): 26-27.

Figure 22: *Spil en Spoel* (1965) magazine cover showing a textile worker in the shadows gazing wistfully over the parking area. Photo by E. Meyling (Source: International Institute for Social History)

democratization, economic progress, and individual freedom followed an international trend. Mobility was the cornerstone of a widely held dream of a society growing in prosperity and individual freedom during the Cold War. Across the U.S. and Europe, planners and (social democrat) politicians agreed the future belonged to the car—and upwardly mobile workers deserved a piece of that future. In the Netherlands, Dutch Labor party leader Joop den Uyl articulated the widely held sentiment among labor leaders by claiming that "everyone has the right to own a car, also the workers," advancing to the material emancipation of the working class.[18]

Dutch automobility became widespread later than elsewhere in the west, however, in 1963, the Dutch government deregulated the guided wage policy that had kept wages low to restore the postwar economy. Only then could workers' wages, disposable income, and consumption levels increase. Combined with dropping production costs and prices of cars—in 1960 the cost of a Volkswagen Beetle equaled the median household income, while a decade later this ratio had fallen to a quarter—car ownership and usage soared. In the period 1960-1970, the number of cars in the Netherlands increased fivefold. In 1960, one out of twenty-three people owned a car, in 1970 this ratio was one in five—similar to surrounding countries. The car became a major item of consumption for families.[19]

Until 1970, however, most working-class car owners did not commute by car. If they could afford private cars, families used them mainly for recreational purposes, as automobile historians Gijs Mom and Peter-Eloy Staal remind us. To put it differently: income did not predict whether families owned a car, but did predict how much they used it. Whereas by 1962 two-thirds of Dutch car owners earned less than 10,000 guilders a year, those earning more than 20,000 drove by far the most kilometers. For lower income groups, private cars—often older, second-hand models—were still mostly used for recreation: the Sunday drive and vacations.[20]

18 "Iedere arbeider zijn eigen auto!," Mobiliteits Museum, accessed June 18, 2021, https://www.mobiel-erfgoed.nl/VMM_NOM_SenS_Auto_iedereen.htm; Ruth Oldenziel and Adri Albert de la Bruhèze, "Contested Spaces: Bicycle Lanes in Urban Europe, 1900-1995," *Transfers* 1, no. 2 (2011): 29-49, here 38; Tim Verlaan, "Mobilization of the Masses: Dutch Planners, Local Politics, and the Threat of the Motor Age 1960-1980," *Journal of Urban History*, 47, no. 1 (October 2019): 136-156, here 14-15.

19 Mark van den Heuvel, "Verschuivende contexten van vrije tijd en vermaak," in *Vertrouwde patronen, nieuwe dromen: Nederland naar een modern industriële samenleving 1948-1973*, ed. Hélène Vossen, Marjan Schwegman, and Peter Wester (IJsselstein: VGN, 1992), 97-100; Gijs Mom, Johan Schot, and Peter-Eloy Staal, "Werken aan mobiliteit: de inburgering van de auto," in *Techniek in Nederland in de twintigste eeuw: Transport, communicatie*, edited by Harry Lintsen and Johan Schot (Zutphen: Walburgpers, 2002), 45-73, here 70; Peter-Eloy Staal, *Automobilisme in Nederland: Een geschiedenis van gebruik, misbruik en nut* (Zutphen: Walburgpers, 2003), 118; John Urry, "The 'System' of Automobility," *Theory, Culture & Society* 21, no. 4/5 (2004): 26, 28-29; Verlaan, "Mobilization of the Masses," 4-5. Car ownership ratio was 2.3 in the United States, 4.8 in the United Kingdom, 4.6 in West-Germany, and 5.3 in Belgium.

20 Mom, Schot, and Staal, "Werken aan mobiliteit," 70; Staal, *Automobilisme in Nederland*, 86-87, 118.

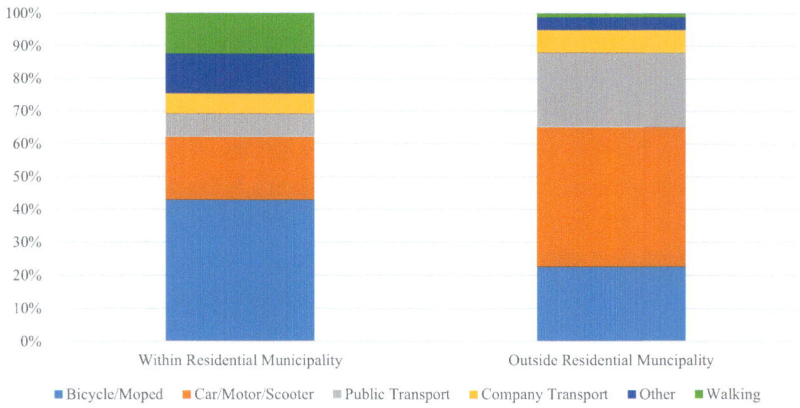

Figure 23: National commuter modal split 1971 (Source: Census 1971)

Most low-income workers still found the total cost of car ownership too high. A 1965 *Spil en Spoel* survey among the few car-owning textile workers revealed how many pooled the costs with extended family. A 21-year-old unmarried worker at cotton spinning factory Bamshoeve in Enschede, bought a second-hand Citroën together with his brother from their savings. To keep costs low, they did small repairs themselves: for insurance, tax, and fuel they paid 25 guilders a week. A 21-year-old woman, employed as office worker at spinning factory Twenthe in Almelo, had bought a Renault together with her brothers and father. They saved a fixed sum per month for repairs, tax, and other expenses, and "whoever has the car pays for the gas." She could not have paid all the costs of ownership herself: "It mounts up! If I had to maintain it myself, it would have to go!" Despite car ownership becoming more common for textile workers, *Spil en Spoel* concluded that "in no way could they afford 'it' from their normal wages."[21] This applied even more so for migrant workers, 1968 statistics showed: 24 percent of Italian workers owned a car versus 14 percent of Spanish workers, and 3 percent of Turkish and Moroccan immigrants—lower shares likely due to Dutch banks' reluctance to provide loans to single households. And like their Dutch counterparts, migrant workers in the 1960s generally did not use their car for commuting.[22]

21 "Een eigen auto," 10-11.

22 Nederlandse Stichting voor Statistiek, *De buitenlandse arbeider in Nederland* (The Hague: Staatsuitgeverij, 1971), 245; "Collectie Yibar: Enschede 1974," IISG Historisch Beeldarchief Migranten, accessed June 18, 2021, http://www.iisg.nl/hbm/toonfoto.php?onderdeel=4&advsearch=1&foto=29&trefwoord=vervoer.

Over time, car commutes became more common. Between 1960 and 1970, the total kilometers traveled by car commuters grew by 500 percent. In the Netherlands, especially among interlocal commuters, the car ranked high, according to the 1971 census (fig. 23): car drivers made up 43 percent, followed by cyclists (23 percent), and public transit users (23 percent).[23] In 1970, 52 percent of Enschede's workers drove to work.[24] Frank Veraart, using CBS 1982 data, observes a similar trend in (former) mining towns. In Heerlen and Kerkrade, respectively 42 percent and 35 percent of interlocal commuters went by car.[25] According to historian Bouwens, Dutch cross-border workers also switched to driving to their German factory or construction jobs, often carpooling with co-workers.[26] By 1978, just over half of Dutch commuters drove to work by car. The other half commuted by other modes.[27] These modal shares suggest that private cars provided a growing number of workers a commuting option. Whether this was a free choice, they did so in an environment with few alternatives.

Car boosters framed automobility as a form of freedom. Others were more critical. Workers commuted by car often out of necessity rather than luxury. This weighed heavily on their household budget. Today's mobility poverty scholars see this as a form of forced car ownership.[28] Some contemporaries agreed. With the wage explosions in 1963, rents and housing prices also peaked. In 1971, a reader of *De Tijd* reflected on the high cost of living in the Netherlands, further aggravated by daily car usage. In addition to a monthly loan repayment of 200 guilders (for two years), car owners had to pay insurance, road tax, fuel, and small repairs, amounting to another monthly 125 guilders, so a total cost of at least 325 guilders to own a car. Like

23 Centraal Bureau voor de Statistiek, *14e Algemene volkstelling annex woningtelling, 28 februari 1971, Deel 7B: Woon-werkverkeer : voornaamste uitkomsten per gemeente* (The Hague: Staatsuitgeverij, 1982), 34-35; Luuk Schreven, *Documentatierapport Volkstelling 1971* (The Hague; Centraal Bureau voor de Statistiek, 2004), 175.

24 Adri Albert de la Bruhèze, "Enschede: An Experiment in Cycling," in *Cycling Cities: The European Experience*, ed. Ruth Oldenziel et al. (Eindhoven: Foundation for the History of Technology, 2016): 41-51, here 43.

25 Frank Veraart and Manuel Stoffers, "Southeast Limburg: Cycling Goes Downhill," in *Cycling Cities: The European Experience*, ed. Oldenziel et al., 65-75, here 71; Frank Veraart, "Cycling in Numbers: 1920-2015," in *Cycling Cities: The European Experience*, ed. Oldenziel et al., 200-229, here 210, 212, 214.

26 Bouwens, *Over de streep*, 71, 119.

27 H. Katteler, W.F. de Heer, and J. Kropman, *Het gebruik van de fiets in Nederland: Een aanvullend onderzoek in de provincies Groningen, Friesland, Zuid-Holland, Zeeland, Noord-Brabant en Limburg* (Nijmegen: ITS, 1978), 63-66.

28 Tobias Kuttler and Massimo Moraglio, *Re-Thinking Mobility Poverty: Understanding Users' Geographies, Backgrounds and Aptitudes* (London/New York: Routledge, 2020), 16.

in surrounding countries, this was a third of the typical 1000-guilder manual workers' monthly income.[29] Employers also voiced concern. Employers in Rotterdam's new Botlek-Europoort industrial area claimed that their workers needed a car to reach isolated port industries, but they considered—citing the Transport Advice Group (TRAG) 1967-1968 survey—the cost of driving a car too high. Company travel cost reimbursements hardly covered the real costs of commuting by car. In many cases—no quantitative data available TRAG noted—cars were purchased "'out of necessity' by people who, for whatever reason, could not afford one." Commuting 40,000 km by car per year, put "a huge burden on the household budget."[30]

Company travel allowances covered some driving costs, but not all. By 1990, trade unions, who had become more involved in workers' mobility, had managed to include in most collective labor agreements in the Netherlands a section on travel cost reimbursements, typically based on public transit fares (Dutch Railways tariffs) or an average reimbursement of ten cents per travelled kilometer. Either way, the first ten kilometers were usually not reimbursed. Within this travel distance, workers paid the full cost. The maximum reimbursement was set at 30 to 40 km. Above that threshold, workers had to pay a large amount out of their own pockets. This policy particularly burdened people forced to commute long distances. On a commuting distance of 60 km, a car commuter had to pay almost 5,000 guilders a year, fivefold a public transit commuter.[31] These workers paid the highest price in an environment with few to no affordable alternatives.

8.3 Accessibility Crisis for the Car-less

In response to increasing unemployment and public spending on benefits, the government extended the "suitable work" requirements of the 1949 Unemployment Act (Werkloosheid Wet): jobseekers had to accept jobs at greater travel distances. During postwar eras of economic growth and full employment, this had not been of great concern for most blue-collar workers. There was plenty of work. If they could not get there by bicycle

29 "Kortweg," *De Kampioen* 86, no. 10 (October 1971): 580; H. van Nooijen, "Een eigen huis," *De Tijd*, 30 June 1971, 5. In comparison, ANWB magazine *De Kampioen* (1971) wrote that a British manual worker spent on average 320 guilders per month on their car with a gross income of 1250 guilders. A German worker spent a third of his income on a private car and a Belgian worker used a quarter of his income.

30 NA, Archive no. 3.17.17.04, Inv. no. 1734, Transport Advies Groep, 'Rapport', 9.

31 Tuinstra, *Woon-werkverkeer in schonere banen*, 16-17.

Figure 24: Philips used the "information buses" shown in this leaflet to recruit rural jobseekers living further afield who had no mobility options to reach Eindhoven. 1964-1970 (Source: Philips Company Archives)

or moped, employers organized buses. This trend was now a thing of the past. Nonetheless, and in contrast to surrounding countries, Dutch law did not consider mobility barriers as a reason for refusing a job offer because, as law professor Jef Langendonck assumed, the Netherlands was a densely populated country with "very well-developed public transit facilities."[32] This was a misjudgment, since in reality, public transit clearly fell short, especially for interlocal commuting.

"Immobility increases unemployment," the Catholic Trade Union claimed in catholic newspaper *De Tijd* (1972). Just making public transit efficient and affordable would not solve unemployment, but would certainly remove one barrier to land jobs, the Union argued.[33] Employers in the isolated Botlek-Europoort area similarly feared car-less jobseekers were unable to get to job interviews.[34] Philips had similar fears and therefore decided in 1964 to run "information buses" ("informatiewagen") on Saturdays for recruiting isolated rural jobseekers (fig. 24). In the following years, the Ministry of Social Affairs launched similar initiatives to solve the transport mismatch for jobseekers. It introduced travel allowances ("Bijdrage-regeling Verplaatsingskosten") for jobseekers based on the cheapest public transit fare and ran mobile regional employment offices ("GABs-on-wheels") in underserved rural regions, similar to Philips information buses.[35] These solutions, however, did not solve the structural problems of eroding mobility alternatives for the car-less.

Cars may have been celebrated as vehicles of democratization, but the promise was only tenable by ignoring those unable to drive cars. Amid

32 NA, Archive no. 2.15.62, Inv. no. 2199, *"Handleiding Passende Arbeid"* (1977), C.R.v.B., 27 March 1975, R.S.V. 1975 no. 341; Wetenschappelijke Raad voor het Regeringsbeleid, *Rapporten aan de Regering: Vernieuwingen in het arbeidsbestel* (The Hague: Staatsuitgeverij, 1981), 94; Jef van Langendonck, *Passende arbeid: Een verkenning in de wetgeving van een aantal Europese landen*, Sociale Zekerheidsreeks (Deventer: Kluwer, 1980), 20-21, 45, 61-62, 66-67, 82-83, 90-91, 112-113; Lonneke Reijrink, "De betekenis van het begrip passende arbeid voor werklozen, kortdurende en langdurige arbeidsongeschikten" (MA Thesis, University of Tilburg, 2010), 11-12, 16, 58.

33 H.S., "Immobiliteit maakt werkloosheid groter," *De Tijd*, 7 November 1972, 11.

34 NA, Archive No. 3.17.17.04, Inv. no. 1734, Transport Advies Groep, 'Rapport,' 3-4, 7-10, 17-18. TRAG noted that Botlek-Europoort workers took action themselves. Some workers used the poorly connected public transit, others rented small buses with co-workers for commuting, but most workers who used "their own transport" travelled by private car or carpooled.

35 NA, Archive no. 2.15.62, Directoraat-Generaal voor de Arbeidsvoorziening van het Ministerie van Sociale Zaken en Werkgelegenheid, Inventory no. 644, Stukken betreffende de aanschaf, gebruik en evaluatie van het gebruik van vier mobiele Arbeidsbureaus (ARBVO-bussen); "Arbeidsbureau op wielen loopt gesmeerd," *Het Vrije Volk*, 14 January 1976, 17; "Regeling voor werklozen: Verhuiskosten vergoed bij nieuwe betrekking," *De Volkskrant*, 22 June 1977, 6. GAB stands for Gewestelijke Arbeidsbureaus.

a car-oriented public discourse, scholars exposed the social risks of car-centered transportation and land-use planning. Economists and geographers in the U.K., U.S., and Sweden observed growing mobility inequalities between people with and without a car. Among the first was economist John F. Kain, who in 1968 coined the "spatial mismatch" concept in an American study on black inner-city jobseekers without cars and public transit.[36] In *The Latent Demand for Urban Transportation* (1968), Lester Hoel et al. highlighted the wider existence of "captive" public transit users, who "must use public transportation regardless of cost or level of service ... in those sectors of the city and at those times when public transportation is provided."[37] International scholarship emerged in the following years, showing this was not a unique American phenomenon. In Sweden, geographer Torsten Hägerstrand revealed that "non-drivers" experienced growing difficulties in "carrying out even modest activity sequences."[38] Peter Wilmott showed that in London's metropolitan area, 42 percent of the 1,928 surveyed people belonged to a car-less household, four-fifths earning less than 500 pounds, two-thirds were unskilled workers and jobseekers who consequently had "less choice of jobs."[39] British town planner Mayer Hillman pointed out that the car-less depended on walking, cycling, and public transit—options which "steadily declined in quality, convenience, and availability" with rising car use.[40]

The Netherlands voiced similar concerns. Already in 1972, Michel van Hulten, progressive Political Party of Radicals (Politieke Partij Radikalen) senator and later state secretary of Transport (1973-1977), warned that car-centrism steered the Dutch mobility system "full speed down a one-way street." Against the backdrop of a growing anti-car counterculture, the politician took up the case for car-less people's mobility needs. He argued that emphasizing ownership growth obscured the fact that about a quarter of all Dutch households were car-less. Even in car-owning households, the family car was by default used by male breadwinners during weekdays, leaving other householders without a car. He claimed that "a society steering

36 John Kain, "Housing Segregation, Negro Employment, and Metropolitan Decentralization," *Quarterly Journal of Economics* 82, no. 2 (1968): 175-197.

37 L.A. Hoel et al., *Latent Demand for Urban Transportation* (Pittsburg, PA: Carnegie Mellon University, 1968), 2-3.

38 Torsten Hägerstrand, *The Impact of Transport on the Quality of Life* (Lund: Lunds Universitets Kulturgeografiska Institution, 1974), 50.

39 Peter Willmott, "Car Ownership in the London Metropolitan Region," *Greater London Council Intelligence Unit Quarterly Bulletin*, no. 23 (June 1973): 5-19, here 11-12, 19.

40 Mayer Hillman, "Social Goals for Transport Policy" (paper presented at the Transport for Society Conference, London, 1975), 13-14. See for a literature review: Hans Jeekel, *Inclusive Transport: Fighting Involuntary Transport Disadvantages* (Cambridge, MA: Elsevier, 2018).

towards full-blown car-orientation" not only ran the risk of an automobility lock-in, but also of socially excluding people, creating a "mobility-proletariat." Not the wage workers in Marxist lexicon, but "the young people, the older people, the women and the handicapped ... who do not have a car at their disposal."[41] Others confirmed the problems sketched by Van Hulten.

In 1976, sociologist Enne de Boer, lecturer at Delft University of Technology, translated international scholarship into a Dutch context and researched the local situation. Like Van Hulten, De Boer showed that the idea of mass motorization advertised by car boosters, was fiction, also in the Netherlands. Having delved into data on car ownership and driver's licenses, he estimated the percentage of Dutch car-less people was between 25 and 40 percent of the total population. Two-thirds of women were car-less and almost three quarters of car-less households were in the lowest income category (up to 1,300 guilders a month). For these people, car-centered transportation and land-use planning led to an imminent "accessibility crisis" ("bereikbaarheidscrisis"): meaning they could capitalize on far fewer opportunities, including jobs.[42]

This accessibility crisis was aggravated by the spatial unbundling of living and working. According to sociologist John Urry based on U.S. and U.K data, decades of car-centered planning had led to what he called "'unbundled' territorialities of home, work, business and leisure that historically were closely integrated."[43] Indeed, as described in previous chapters, company buses had partly enabled this separation between countryside living and urban working in the Netherlands too. In the 1960s, spatial unbundling became deliberate government policy too.

41 Michel van Hulten, *'Gratis' openbaar vervoer* (Deventer: Kluwer, 1972), 19, 37-39.

42 Boer, "Mobiel en niet-mobiel," 48, 56-60; A.I.J.M. van der Hoorn and M.R. Mulder, "Een gedisaggregeerde registratie van de ontwikkeling van rijbewijs- en autobezit," (The Hague: Dienst Verkeerskunde, Rijkswaterstaat, 1981), 7-9. In general, possession of driver's licenses and cars correlated with income class, gender, and geographic location. A 1981 study by Rijkswaterstaat about car ownership and driver's license possession, included data about car ownership among people with a driver's license set against income class and gender. The data indicated that access to private cars was significantly lower among women (33 percent) than among men (90 percent). In the income class up to 1,300 guilders circa forty percent of men owned a car, against eleven percent of women; income class up to 2,400 guilders 76 percent of men owned a car and forty percent of women; and from incomes of 2,400 more than ninety percent of men owned a car against little more than half the women in this income class. Driver's license possession was highest in income group of 2,400 guilders a month (94 percent men, 68 percent women) and lowest in the 1,300-income class (fifty percent men, 39 percent women). Having a driver's license, however, did not automatically imply access to private cars, since the one family car was typically used by male breadwinners one working days.

43 Urry, "The 'System' of Automobility," 28, 30.

Central planning had characterized Dutch postwar governance: from the nationwide guided wage policy and social security to centrally planned infrastructure development and land reclamation for agriculture, housing, and industries.[44] Spatial distribution was a central element of this Dutch political culture. The Dutch government feared an uneven distribution of the population. Population growth and lack of space were already taking problematic forms in the Randstad conurbation of Amsterdam, The Hague, Rotterdam, and Utrecht.[45] The Ministry of Physical Planning proposed in 1966 to "bundle" the population into designated growth nucleuses and relocate industrial and business areas, effectuated in the Second Spatial Planning Bill (Tweede Nota voor Ruimtelijke Ordening). This was an attempt to meet the growing demand for housing, pushed by the postwar baby boom, while guiding associated urbanization by designating sixteen nucleuses where growth was allowed. The result was an extensive—and in the Netherlands unprecedented—suburbanization of upwardly mobile households seeking more living space and comfort.[46]

In the Second Spatial Planning Bill, the government presented automobility as a dynamic exponent of modern Dutch society. The private car, symbolizing collective welfare and individual freedom of movement, was also the vehicle that should enable spatial distribution.[47] In effect, this

44 Gijs Mom, "The Emancipation of the Urban View: Dutch Spatial Planning in an International Context (1920-1950)," in *Builders and Planners: A History of Land-use and Infrastructure Planning in the Netherlands*, edited by Jos Arts et al. (The Hague: Eburon Academic Publishers, 2016), 73-139, here 127-136.

45 Kees Schuyt and Ed Taverne, *1950: Welvaart in zwart-wit* (The Hague: Sdu Uitgevers, 2000), 104-105.

46 Boer, "Mobiel en niet-mobiel," 48, 56-60; Greet De Block, David de Kool, and Bruno de Meulder, "Paradise Regained?", in *Builders and Planners: A History of Land-use and Infrastructure Planning in the Netherlands*, edited by Jos Arts et al. (The Hague: Eburon Academic Publishers, 2016), 17-68, here 61-63; Michael Gommers and Piet Bovy, *Vervoermiddelgebruik bij woonforensen uit kleine kernen*, ed. Planologie en Architectuur Onderzoeksinstituut voor Stedebouw (Delft: Delftse Universitaire Pers, 1988), 10-11, 13; Len de Klerk and Ries van der Wouden, *Ruimtelijke ordening: Geschiedenis van de stedelijke en regionale planning in Nederland, 1200-nu* (Rotterdam: NAI010, 2021), 18; Ministerie van Verkeer en Waterstaat, *Structuurschema Verkeer en Vervoer* (The Hague: Tweede Kamer der Staten-Generaal, 1976-1977), 122; Ministerie van Verkeer en Waterstaat, *Structuurschema Verkeer en Vervoer: Deel C, Adviezen van de Raad van de Waterstaat en van de Raad van Advies voor de Ruimtelijke Ordening* (The Hague: Staatsuitgeverij, 1978-1979), 97; Ed Nozeman, "Groeikernen: reisafstanden moesten kleiner worden maar het omgekeerde gebeurde," *ROM-Magazine*, no. 6 (June 1987): 8-12; Verlaan, "Mobilization of the Masses," 2, 6; Ries van der Wouden, "Succes of falen? Een halve eeuw verstedelijkingsbeleid in Nederland," *Ruimte & Maatschappij* 8, no. 1 (2016): 10-11, 13.

47 Schuyt and Taverne, *1950: Welvaart in zwart-wit*, 171.

policy increased the necessity to travel greater distances by car, as De Boer warned.[48] Moving jobs away from where people lived, however, had been set in motion decades earlier.

Decentralization of industrial workplaces had been government and company policy since the early twentieth century—at the time with rail-mobility in mind, in the 1970s with automobility as guiding principle. Dutch suburbanization was modest compared to the American development path, and in *De Nederlandse industrie: Vernieuwing, verwevenheid en spreiding,* Oedzge Atzema and Egbert Wever highlight the trend was still significant. Whereas industries initially located in (urban) areas with large unskilled labor reserves, they increasingly (re)located to new industrial areas in the urban outskirts, prompted by business economic motives—as detailed by Gerardus Delfgaauw in 1932. In the postwar era, heavy manufacturing and wholesale industries migrated to urban outskirts.[49]

Car-centered suburbanization led to the emergence of car-only environments. In the port of Rotterdam, the newly westward developing Botlek-Europoort was poorly accessible for other modes than private cars. In 1967-1968, the Transit Advice Group TRAG described it as an isolated area with "little infrastructure and an almost total lack of public transit" far from the region's main residential areas, unlocked by roads and the new Botlek-tunnel for cars only.[50] Similarly, with the increasing traffic of Hoogovens commuters, from 4,300 in 1962 to 11,300 in 1975, roads around the blast furnaces became congested. The proposed solution: building more roads and tunnels for cars.[51] Extensive studies by Frans Messing and Hans Kasper et al. on Dutch mine closures show that in government restructuring plans, the idea was: jobs had to be created without considering their location or how people should get there. Unemployed miners living in the former mining region—as good as half the mining population—were forced to commute to workplaces farther afield. Accommodating these commuters,

48 *Structuurschema Verkeer en Vervoer*, 122; *Structuurschema Verkeer en Vervoer: Deel C*, 97; Boer, "Mobiel en niet-mobiel," 48, 56-60; Gommers and Bovy, *Vervoermiddelgebruik bij woonforensen uit kleine kernen*, 10-11, 13; Nozeman, "Groeikernen: reisafstanden," 8-12; Verlaan, "Mobilization of the Masses," 2, 6; Wouden, "Succes of falen?," 10-11, 13.

49 Oedzge Atzema and Egbert Wever, *De Nederlandse industrie: Vernieuwing, verwevenheid en spreiding* (Assen: Van Gorcum, 1999), 97; Gerardus Delfgaauw, *De tendenzen tot decentralisatie in de vestiging der nijverheid*, Serie voor Stedebouw VII (Amsterdam: Nederlandsch Instituut voor Volkshuisvesting en Stedebouw, 1932).

50 NA, Archive no. 3.17.17.04, Inv. no. 1734, Transport Advies Groep, 'Rapport,' 1-2, 20, 35.

51 Bram Bouwens et al., *Door staal gedreven: Van Hoogovens tot Tata Steel, 1918-2018* (Bussum: Uitgeverij Thoth, 2018), 119-120.

the government spent a total of 169,500,000 guilders on infrastructural improvements with car accessibility the basic tenet.[52]

And so, bicycles remained important for many local commuters. Like today, the bicycle was particularly indispensable for low-income, car-less people.[53] The 1971 census revealed that almost half the Dutch local commuters rode bicycles or mopeds, followed by cars, walking, and public transit. In Enschede, 60 percent of local workers and jobseekers travelled to jobs by bicycle or moped in 1970, Adri Albert de la Bruhèze shows.[54] And this was the general pattern the Institute for Applied Sociology (Instituut voor Toegepaste Sociologie) showed in *Het gebruik van de fiets in Nederland* (1978)—notwithstanding local differences. Based on almost a thousand surveys among people between 15 and 74 years old, the report showed most bicycle commuters lived within a 2-6 km radius from their work. A fifth of cycling commuters claimed to cycle out of necessity ("noodzaak"), because their workplace was too far to walk, and alternatives were non-existent.[55] The newly established Dutch Cyclists' Union—representing urban cyclists through protest actions and lobbying—stood up for these cycling commuters. In 1978, the Union claimed that "it is not by chance that the weakest in society are also the 'captive' cyclists and therefore automatically also the weakest in traffic," referring to cyclists that had no choice but to travel by bicycle. With growing car usage, cyclists' position on the road became further contested as cars take up lots of space by virtue of their size and speed, and the government's investments in their infrastructures. The Cyclists' Union too was concerned about transport poverty without investments in safe pathways for the many people that relied on their bicycle for the daily commute.[56]

52 Hans Kasper et al., *Na de mijnsluiting in Zuid-Limburg: 35 jaar herstructurering en reconversie 1965-2000 en een doorkijk naar 2010* (Maastricht: Stichting Behoud Mijnhistorie, 2013), 12, 14, 39, 59, 72, 77, 95, 121-122; Frans Messing, *Geschiedenis van de mijnsluiting in Limburg: Noodzaak en lotgevallen van een regionale herstructurering 1955-1975* (Leiden: Martinus Nijhoff, 1988), 192-193. Highway connections were improved between Maastricht and Heerlen (A79), towards industrial and growth centers in Belgium and Germany.

53 Karel Martens, "Role of the Bicycle in the Limitation of Transport Poverty in the Netherlands," *Journal of the Transportation Research Board* 2387, no. 1 (2013): 20-25.

54 Albert de la Bruhèze, "Enschede: An Experiment in Cycling," 43.

55 Katteler, Heer, and Kropman, *Het gebruik van de fiets in Nederland: Een aanvullend onderzoek in de provincies Groningen, Friesland, Zuid-Holland, Zeeland, Noord-Brabant en Limburg*, 63, 65-66, 71, 94, 118, 124; Jan Ploeger, "Nederlandse fietsers neergelegd in cijfers," *Vogelvrije Fietser* 3, no. 4 (1978): 19-20.

56 International Institute for Social History, Archive no. ARCH01969, Archief ENFB, 1975-1997, Inv. no. 1, Minutes 5 April 1978, 1-3. Among the few contributions dealing directly or indirectly with transport poverty were: Anita Boelsums and Paulien Osse, "Bevrijdde het rijwiel

Unlike in the interwar and first postwar decades, the bicycle was less commonly used for longer distances. Generally, distant jobs had become poorly accessible without a car. New job locations, like those in the Europoort-Botlek area were poorly unlocked.[57] Nation-wide, at least a third of all workers could not get to their workplace by public transit, 1978 statistics indicated.[58] Little more than a decade later, labor union FNV claimed that "hardly any industrial site is easy to reach by public transit." The absence of public transit services or nearby stops made driving an absolute necessity rather than a luxury for many commuters.[59] This was the result of political choices, favoring automobility.

Making way for cars came at the expense of public transit. Between 1955 and 1975, the bus, tram, and train share of commuters' total travelled kilometers dropped from 53 percent in 1955 to 18 percent in 1975. In places like Eindhoven, declining passenger numbers and the absence of municipal subsidies, had led to Eindhoven's bus operator going bankrupt in 1971.[60] Despite this downward trend, the bus was still an important and affordable alternative to bicycles and cars for interlocal commuters. According to the 1970 census, three percent of local workers in Enschede used public transit; over a fifth of interlocal commuters still travelled by bus. Similarly, in (former) mining towns Heerlen and Kerkrade, respectively 16 and 25 percent of workers employed outside these towns commuted by bus.[61] Despite a significant share of interlocal commuters using public buses, people who relied on buses were confronted with eroding bus services in the following decades. This was not only the result of people's shifting modal preferences, but also political decisions.

National passenger transport depended on a concession-system. Moreover, public bus companies suffered from severe government cutbacks in public transit in the 1970s. The Dutch Consumers' Union claimed in 1978

de vrouw," *Vogelvrije Fietser* 2, no. 5 (1977): 18-19; Enne de Boer, "Het onrecht van de sterkste: Sociale ongelijkheid in verkeer en vervoer," *Vogelvrije Fietser* 5, no. 3 (1980): 18-19, 28; Charles Crombach, "Gelijke vervoersmogelijkheden voor iederen," *Vogelvrije Fietser* 4, no. 2 (1979): 10-11.

57 NA, Archive No. 3.17.17.04, Inv. no. 1734, Transport Advies Groep, 'Rapport,' 3-4, 7-10.

58 Centraal Bureau voor de Statistiek, *De mobiliteit van de Nederlandse bevolking in 1978* (The Hague: Staatsuitgeverij, 1981), 32.

59 Tuinstra, *Woon-werkverkeer in schonere banen*, 31-32.

60 Schreven, *Documentatierapport Volkstelling 1971*, 175; Staal, *Automobilisme in Nederland*, 118; Frank Veraart, Martin Emanuel, and Ruth Oldenziel, "Eindhoven: Engineering a Path for Bikes? Eindhoven," in *Cycling Cities: The European Experience*, ed. Oldenziel et al., 53-63, here 53, 57; Verlaan, "Mobilization of the Masses," 4-5.

61 Albert de la Bruhèze, "Enschede: An Experiment in Cycling," 43; Veraart, "Cycling in Numbers: 1920-2015," 210, 212, 214.

that public bus services fell short for greater travel distances, not living up to its function as mobility alternative to automobility. Around 800 of 3,300 residential areas ("woonkernen") had no public transit connections, especially in sparsely populated rural areas. But even in villages *with* a public bus connection, the frequency—less than two buses per hour—was often so low "that you could hardly speak of any public transport at all," the Consumers' Union argued.[62] In *Gemiste Bussen* (1982), ROVER—a national advocate for public transit passengers founded in 1971—reached a similar conclusion. ROVER blamed the Dutch government for divesting in (regional) public buses, aggravating already precarious mobility options for car-less people.[63]

The government's outsourcing of public transit to the market was part of the problem. In 1967, as public transit operators' deficits increased, the national government decided to cover these deficits through subsidies. This led to an explosive growth in subsidies. In the 1970s, the Ministry of Transport and Public Works sought ways to make public transit more cost-efficient, especially in sparsely populated rural regions. The government's Multi-year Plan for Passenger Transport (Meerjarenplan voor het Personenvervoer) 1976-1980 provides a glimpse into how the state envisioned helping an ailing public transit sector. The plan, centered around the maintenance and expansion of car infrastructures (in rural areas), prescribed a "reasonable—adjusted as much as possible to demand—public transit provision" as alternative to automobility. Den Uyl's progressive government instigated cutbacks, raised fares, and issued an assessment tool in 1976—the Standardization System Region Transport (Normeringssysteem Voorzieningenniveau Streekvervoer). Below a 5-10 passenger threshold, bus services were to be discontinued. The tool was inscribed with market-based principles: the policymakers assumed the tool would automatically lead to the most efficient distribution of state subsidies without too much government involvement.[64] Pushing public transit into the market would gradually put an end to public buses in sparsely populated areas, ROVER claimed—even though citizens depended on them.[65] ROVER's warning became reality in the following years. Under

62 "Openbaar streekvervoer per bus (deel 2)," *Consumentengids* September (1978): 422-424.
63 Frits Bergsma, "ROVER-hoofdman aan het woord: Sommigen zien ons als lastige pottekijkers," *Vogelvrije Fietser* 4, no. 4 (1979): 22-23; ROVER, *Gemiste Bussen: Een nota van de Werkgroep Streekvervoer van de Vereniging Reizigers Openbaar Vervoer (ROVER) met als onderwerp: Het voorzieningenniveau van het streekvervoer in Nederland*" (Haarlem: Stichting ROVER, 1982), 30.
64 Ministerie van Verkeer en Waterstaat, *Meerjarenplan voor het personenvervoer 1976-1980* (The Hague: Staatsuitgeverij, 1975), 12-13, 31-32.
65 ROVER, *Gemiste Bussen*, 29.

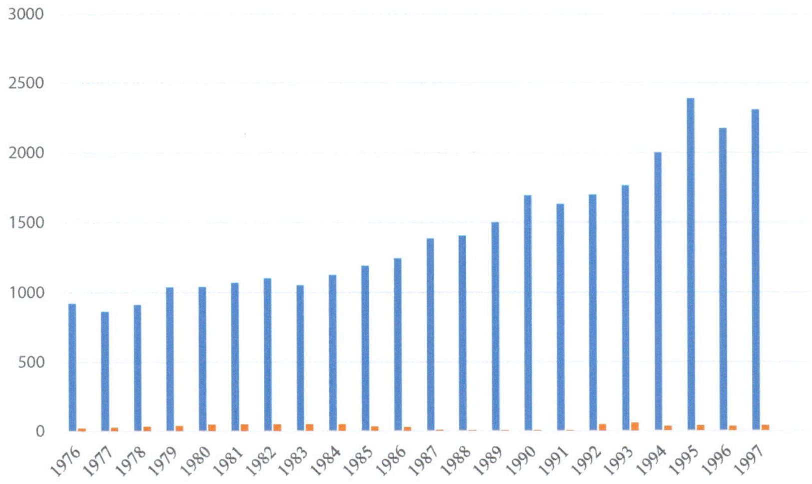

Figure 25: National government spending in millions on cycling infrastructure (orange) and highway construction (blue), 1976-1997 (Source: Dekker 2021, 201)

a conservative, center-right government, the cost price of public transit steered government austerity measures rather than public value.

Following the neoliberal turn after the 1982 elections, the government considered cars the most efficient mobility option, discounting other options as inefficient. The idea that neoliberalism and private automobility are co-travelers, thus also seemed to apply to the Netherlands. Under the right-wing cabinet of Christian-democrats and conservative-liberals, Minister of Transport Neelie Smit-Kroes cut back spending on cycling infrastructure and public transit. Instead, the ministry invested heavily in highway construction (fig. 25). Essentially, since political choices were guided by partial interests, the mobility of some (i.e. car drivers) was enhanced, while that of others was constrained.

The government's role in providing a safety net for underserved groups in society failed, while company welfare and transport programs waned. In the 1980s, in line with academics who had first identified the issue, activists voiced concerns about the consequences of this trend for already marginalized citizens. Women in particular suffered the consequences of this trend, as they generally relied more often on walking, cycling, and public transit, compared to men who drove the family car to work, as feminist urban planner Henriette van Eys observed.[66] The 1982 Domestic Work Project

66 Henriette van Eys, *Rotterdam, een stad voor iedereen? Een onderzoek naar een rechtvaardiger verdeling van ruimte en voorzieningen voor alle stadsbewoners* (Rotterdam: Dienst

Group (Projectgroep Huishoudelijke Arbeid) at Groningen University and the Women on Welfare Committee (Komité Vrouwen in de Bijstand) revealed that spiking public transit costs and welfare cutbacks disproportionally burdened low-income (single) women. They were thrown back on bicycles, "even though their bicycles are in poor condition because there is no money for the repair man," newspaper *Het Parool* reported in 1982.[67] Alongside other social disadvantages, mobility barriers further aggravated these women's social isolation and struggle to capitalize on life opportunities. Being without a car, meant fewer to no job options.

By 1990, under pressure of worsening traffic congestion across the country, the government, union FNV, bicycle working groups, and the Cyclist's Union urged employers to facilitate employees in their commute by encouraging non-car alternatives. FNV urged employers to locate near public transit hubs or lobby for nearby bus stops. Under the banner of the Bicycle Masterplan, a national pilot and research project in the field of cycling, mostly focused on infrastructure, employers were also encouraged to lobby for bicycle path construction to job sites, lighting along remote industrial roads, bicycle parking and repairs, and fiscal benefits for employees when purchasing a bicycle for commuting—interventions that had been common in earlier decades as we have seen. Some employers took up the call. Many did not, a 1993 evaluation showed.[68] In contrast with the first postwar decades, there was no structural support from the government and employers for car-less

Stadsontwikkeling, 1981); Van Eys, *Stedelijke Bereikbaarheid: Een voostel tot kwaliteitsverbetering en kostenbesparing in het verkeers- en vervoerssysteem te Rotterdam* (Rotterdam: Stad en Landschap, 1983); Van Eys, *Emancipatoire aandachtspunten in stedebouwkundige plannen* (Rotterdam: Gemeente Rotterdam, 1986). To date little is known about the mobility of migrant workers in this period. A later report by consultancy bureau MuConsult, observed that compared to Dutch counterparts, people of Surinamese, Turkish, and Moroccan descent more often lived in cities and more often relied on public transit and less often used bicycles for everyday mobility. Moreover, De Boer had marked areas were most immigrants lived coincidentally as risk areas for transport poverty. The fact that Moluccans were now housed in specially allocated neighborhoods in the fringes of villages and cities—closer to activity centers, still isolated—and other low-income citizens with an immigrant background had no option but to reside in cheap homes in deprived inner cities, makes it likely they experienced mobility barriers to access jobs too. MuConsult, "Mobiliteit van allochtonen: Verslag van een voorstudie" (Utrecht: MuConsult, 1994); Boer, "Mobiel en niet-mobiel," 47-78.

67 "'Bijstandsvrouwen' met kinderen redden het niet," *Het Parool*, 23 October 1982, 7.

68 Tuinstra, *Woon-werkverkeer in schonere banen*, 5; Fietsersbond ENFB, *Werken met de fiets* (The Hague, Hoofdafdeling Infrastructuur en Verkeer, Ministerie van Verkeer en Waterstaat, 1992), 14-15, 19-27, 33-34, 42; Stichting Werkgroep 2Duizend, *Brochure: Werken met de fiets: Evaluatie van de ontvangst door de doelgroep, eindrapport*, (Amersfoort: Stichting Werkgroep 2Duizend, 1993).

workers' commute and job accessibility. More recent scholarly observations about the negative correlation between mobility barriers and people's ability to land jobs, illustrate we still struggle with this legacy today.[69]

Conclusion

The 1970s had set in motion a historical transformation from paternalist to neoliberal views on workers' mobility and job accessibility—with fundamental consequences. Carpenter K's case illustrates how closely related (im) mobility and job access are. His case marks a pivotal shift in thinking about who bears responsibility for facilitating workers' commute. Assumptions that guided the extension of suitable work requirements appeared inconsistent with car-less jobseekers' current mobility situation. Coupled with an already short-falling mobility system for car-less people, the state and employers' withdrawal had left them largely to their own devices. Local commuters could cycle, though not seldomly along unsafe routes. In order to reach suburban and remote jobs, car commuting had become unavoidable.

In the 1970s, the everyday commute became a matter of individual workers' responsibility. Not only for people living near jobs, but also for those living further away and with limited mobility options. Whereas historians typically date the shift from patriarchal state and company welfare to neoliberal responsibility in the 1980s, "Leaving Workers to their Own Devices" argues that the 1970s marked a transition towards a neoliberal governance of workers' mobility. We have seen that the Dutch government restricted itself to safeguarding basic preconditions like transport infrastructures and subsidies in the first postwar decades. In short, during the 1970s recession, the Dutch government cut spending on public transit and considered the everyday commute an individual's responsibility. At the same time, while car-less jobseekers like K. looked to employers for solutions, employers—previously key in enabling further afield workers' mobility—withdrew from direct involvement in workers' mobility amid a global wave of deindustrialization set in motion during the 1960s. Consequently, workers had to fend for themselves, raising barriers for some to land jobs.

69 See e.g. Jeroen Bastiaanssen, Karel Martens, and Gert-Jan Polhuijs, "'Geen rijbewijs, geen fiets, geen ov-aansluiting, geen baan': Vervoersarmoede in Rotterdam-Zuid," *Verkeerskunde* 5, no. 5 (2013): 44-50; Jeroen Bastiaanssen, Daniel Johnson, and Karen Lucas, "Does Transport Help People to Gain Employment? A Systemic Review and Meta-Analysis of the Emperical Evidence," *Transport Reviews* 40, no. 5 (2020): 607-628; Karel Martens, Marnix ten Holder, and Jurriën Thijssen, "Vervoersarmoede Bestaat," *Verkeerskunde* 2, no. 11 (2011): 34-38.

Whether and how people travel is not only the result of how available mobility options are distributed over society, as Mimi Sheller emphasizes in *Mobility Justice*.[70] We have seen how power relations and systems of governance enable or disable workers' mobility. "Workers Left to their Own Devices" has shown that workers' mobility and job accessibility are affected by the spatial organization of living and working, their ability to bear the cost of traveling in terms of time and money, and to influence or be represented in decision-making processes. These aspects of workers' mobility had previously been addressed by stakeholders (e.g. employers) facing the problem of mobilizing a labor force. In the 1970s, academics put these themes on the agenda, not as a challenge solely for workers in an industrial society, but as a broader social problem to be solved.

70 Mimi Sheller, *Mobility Justice: The Politics of Movement in an Age of Extremes* (London/New York: Verso, 2018), XV, 9-11, 27-28.

Conclusion

No Bus, No Bicycle, No Job has analyzed how workers' mobility and job accessibility changed over time, and how blue-collar workers, employers, and the government have shaped this change throughout the twentieth century in the Netherlands. By closely examining workers' past mobility options and barriers in different periods, this book has shown that factors considered key in contributing to mobility poverty today are of all times, and have been addressed by various actors in the past, including workers and employers—unrecognized forces in historiography.

Against the background of what historians Gijs Mom and Ruud Filarski coined the "mobility explosion," I have shown that the relative distance, time, and cost of bridging distances, decreased for most workers over the course of the twentieth century. In addition to rail-based modes, bicycles and buses led to greater time-space compression for manual workers and jobseekers during the interwar period, complemented by mopeds and cars in later decades. Not everyone, however, enjoyed the benefits of widening travel horizons. Like elsewhere in Western Europe, new transport technologies—bicycles, buses, mopeds, and cars—became more affordable and available to manual workers. Still, I have shown that these modes were curtailed in various ways by the spatial organization of living and working, mobility barriers, and company politics of control—thereby shying away from a technological determinist take on historic changes in workers' mobility.

Today's mobility scholarship indicates that low-income workers and jobseekers have more limited travel horizons than upper- and middle-class people. They are more often without a private car, cannot find affordable housing near jobs, rely on failing public transit, and have limited network capital to overcome barriers. As we have seen, blue-collar workers also faced these challenges in the past. Since the early-twentieth century, well before the wider implementation of car-centered planning, the absolute distance from workers' homes to industrial-type job locations increased. Prompted by business economic motives, new industries and industrial areas located further from residential areas. Existing industries in urban locations attracted workers further afield. The lack of affordable housing near jobs to

Bek, Patrick, *No Bicycle, No Bus, No Job: The Making of Workers' Mobility in the Netherlands, 1920-1990*. Taylor & Francis Group 2022
DOI: 10.5117/9789463723183_CONCL

bridge the gap was a problem for both workers and managers, forcing workers to commute greater distances. In the postwar era, company managers' encouraged countryside living and urban working in a paternalistic attempt to deproletarize the labor force and ensure a certain standard of living. This spatial mismatch between affordable housing and industrial jobs, however, was aggravated in the following decades by the national policy of spatially unbundling living and working, formalized in the Spatial Planning Bill (1966). This car-centered transportation and land-use planning, adversely affected workers' ability to get to work without a car. In these efforts, the state and employers normalized long-distance commutes.

For shorter commuting distances, the bicycle remained an important mobility option. Though the average distances travelled by bicycle declined: around the time of the 1938 miner bus boycott, cycling up to 40 km to work during the summer months was not uncommon, but the postwar decades saw lower bicycle commuting distances. The 1947 and 1960 census indicated workers within a 12 km radius from jobs most often cycled to work. By the 1970s, this was 6 km. This likely had to do with the availability and affordability of new alternatives for longer distances like mopeds, company buses, and cars. Despite declining cycling shares across Europe and the Netherlands, for Dutch factory workers, cycling remained an important commuting option.

In the period of economic boom between 1947 and 1970, many interlocal commuters to suburbanized (highway) locations used company buses to get to work. Widespread housing shortages combined with rapidly expanding industries and the employers' search for cheap labor, led to an enormous growth in commutes over greater distances. Based on the material presented in *No Bus, No Bicycle, No Job*, I argue that Dutch public transit structurally fell short for such distances. Strict government regulation of timetables and passenger transport routes, coupled with divestments, made public transit a poor alternative for most blue-collar workers. Company-subsidized buses, however, extended the action radius of thousands of rural jobseekers, cross-border workers, and migrant guest workers, groups devoid of affordable alternatives from their isolated housing locations. These buses enabled them to travel to (better paid) jobs and Dutch rural workers could at least remain in their beloved village. In Limburg, buses also replaced bikeable commutes of 4 km, encouraged by mining companies that partly reimbursed bus fares for these distances because of the hilly terrain. In other cases, workers were only reimbursed or even allowed to use company buses for 7 to 12 km distances. In any case, the commuter bus became a widespread phenomenon in postwar Netherlands.

The first two chapters showed that amid economic recession, unionized workers with limited resources displayed organizational and improvisational strength in overcoming mobility barriers. Workers and jobseekers reverted en masse to cycling over more expensive public transit, and in some cases used paratransit bus services set up by local entrepreneurs, co-workers, and employers like Philips. Even when state sanctioned bus regulations led to price hikes, miners actively resisted this top-down decision by boycotting buses and cycling to work instead. This not only illustrates the vital role of bicycles for workers' agency, but also highlights the self-governing qualities of manual workers.

This study, however, also revealed an important shift in workers' ability to decide how to commute. The Second World War was a turning point. Most European citizens struggled with widespread scarcity and getting around. This also affected Dutch workers' ability to get to work, as rubber tire shortages limited bicycle's role as a lifesaver and the few non-confiscated buses had no fuel or spare parts. During the war, the German occupying government and industrial employers intervened to keep the war economy going. Under the umbrella of general austerity, the government drafted accessibility thresholds and set the normative framework for those who deserved support with their mobility and were considered able to fend for themselves. This support took the form of prioritizing workers in vital industries by distributing scarce bicycle tires and spare parts, and extra (company) compensation for higher travel costs or (free) access to emergency buses. Whether someone deserved such support depended on whether they were employed in critical industries for the war economy and geographic location—i.e. proximity to work or public transit hubs. Even though unions had been banned by the occupier, workers in vital industries could somehow negotiate their mobility through their employers. This was a preamble for the postwar decades.

No Bus, No Bicycle, No Job contributes to existing labor and mobility historiography by unravelling and detailing the underlying politics that governed how dockers, miners, and factory workers got to work. Mimi Sheller has argued that precarious mobility is not just a matter of uneven distribution of transport modes over society, but a question of power, privilege, and political representation too. Tracing the locus of control over workers' mobility, I have shown that the ability (and willingness) to govern mobility is prone to social processes of negotiation and changing power relations, interlinked with unpredictable socio-economic developments like industrialization, deindustrialization.

Once the most severe shortages had been resolved in the postwar decades, industrial managers attempted to gain greater control over workers' mobility via different strategies of control and discipline. The gospel of efficiency and profit that guided the interwar management of factory shopfloor and company housing, was now also translated into the domain of mobility outside the factory gate. From the Second World War until the 1970s, industrial employers collaborated with various stakeholders in shaping workers' mobility. This gave them the power to govern workers' mobility according to the company's interests and planning horizons that were more intense than before. Company concerns about workers arriving late for their shifts, not arriving at all, or getting injured in traffic, were genuine expressions of paternalistic concern. These were also expressions of fear about production loss and mounting costs. This ambivalence between care and control and between empowerment and disempowerment was part and parcel of employers' involvement. Like labor historians have shown for shopfloor management and company housing, this book shows these two logics of care and control were not competing or contradictory framings, but shaped each other in paternalistic power relations. In their attempt to reduce mobility barriers for workers, however, they also discursively set the boundaries of desirable and undesirable mobility practices—an expression of employers' paternalist care and ambitions to control.

Under the guise of postwar reconstructionist ideology, modernist visions of an industrial and motorized future, paired with technocratic ambitions of control over people, individual mobility (bicycles) and collective mobility (buses) became subject to company intervention. Mobility—like housing and other social amenities—was rationalized as a proxy for managing resources and became linked to production and economic growth. The daily commute had to be economically efficient—safe, swift, and at low-cost—and thus Dutch managers sought to (re)configure workers' commutes. Those aspects managers considered deviant or unsuitable in light of hampering workers' spatial movements and industrial productivity, were disciplined until they fitted the set norms. The strategies to achieve this differed per mobility option and consequently per category of workers, *No Bus, No Bicycle, No Job* showed.

In controlling and maintaining the material standards of bicycles, buses, and mopeds, employers exercised different levels of control. In the area of private mobility of bicycles and mopeds (peri-)urban workers used, employers implemented a more indirect approach, seeking the help of local traffic police officers and safe traffic advocacy group VVV—who had formed similar car-oriented safety coalitions like in the U.S. and elsewhere in Europe. In the case of collective mobility of buses for rural and migrant workers, employers applied direct, top-down control via contracts with bus companies or setting up company garages.

Most higher-skilled and better paid (peri-)urban workers, who cycled to work until the 1970s, were exposed to company policing and schooling. They were not directly curtailed in their decision how to commute. Unskilled, ununionized, and lower-paid rural and migrant workers, however, generally enjoyed less autonomy in determining how and when to move, as they lived in places devoid of mobility alternatives and thus heavily relied on company support. Even though these buses granted them affordable and fast mobility options that they otherwise might not secure, dependence on company buses also forced them to use a mobility option that only existed as long it fitted their employer's planning horizon.

Company buses filled a lacuna in the Dutch mobility system. Employers had attempted to organize buses to circumvent the 1938 "autobus wars," avoid the 1940s mobility barriers, and overcome failing public transit. This company-dependent system, however, was not structural for the long run, but depended on economic expansion when employers faced labor shortages. This socio-technical system was based on a network of social relations that centered around industrial growth, making it vulnerable to global economic fluctuations. Following the global wave of deindustrialization and the 1970s economic recession, company buses disappeared in Limburg's mining region and Twente's textile region, and were downsized in other places. Moreover, rather than direct interventions, employers switched to more indirect policies like travel cost reimbursements. It remains difficult to assess how many blue-collar workers and jobseekers consequently experienced barriers to access jobs, but wider concerns about the more limited travel horizons of those who did not own a car at the time suggest that (car-less) jobseekers like carpenter K. suffered the consequences.

This book shows that if represented by powerful actors like industrial companies, people's options to travel were more likely to be secured. This book has, however, also shown that a focus on state and company power alone cannot fully explain historic change. Deviating from a strict Foucauldian perspective that denies individual agency, I presented ample evidence that blue-collar workers also shaped the everyday commute through processes of negotiation and bargaining. The role of trade unions in representing workers seemed limited for solving mobility challenges—with the exception of the mining region, where unions backed the 1938 bus boycott and later sat permanently on the Miner Transportation Committee. Yet even without such formal representation, workers' mobility was often the outcome of self-governing processes in which they participated actively. At different times, manual workers resisted a status quo that was not in their favor by repurposing bicycles, buses, mopeds, and cars for utilitarian use, and protested, negotiated, and bargained for accessible mobility options. The 1970s, however, marked a turning point for many low-income workers.

No Bus, No Bicycle, No Job concludes with a historical shift from paternalistic company influence over workers' travel to neoliberal public governance, where the responsibility for being mobile was placed with workers and jobseekers, rather than the state and big business. Again, workers had to fend for themselves like during the interwar recession. Mobility barriers had been similar for most low-income workers and jobseekers earlier. By the 1970s, like observers in the U.K., U.S. and Sweden agendized at the time, this book showed that the ability to get to work had become more socially differentiated in the Netherlands too. Those who could afford a car were able to access distant jobs, but car-less immigrants, women, and younger jobseekers like carpenter K. struggled to access remote jobs. The Dutch government initially tried to mitigate 'transport poverty' for jobseekers, but generally failed as a shield for underserved groups by subsidizing automobility and divesting in cycling and public transit.

The five different cases studies over a period of sixty years following different historical actors and transport modes combine perspectives of mobility and labor history. They offer insights in workers' everyday commute. Still, questions remain. First, the question of workers' agency has been explored throughout the book in various ways: from modal choice to protest, petitions, and voicing complaints to managers. For the postwar period, however, workers' experiences and field of actions deserve further study. Especially for company disciplining of workers, in chapters 6 and 7, inspired by Michel Foucault's work on power and discipline, I focused on the social construction of workers as mobile subjects, rather than studying workers' resistance or experiences. This was partly due to the limits of what I could distill from available sources, and partly due to my focus on power and company control. A methodological shift towards working with ego documents and oral history might enable insight in how such disciplinary interventions impacted workers. Second, geographical information (GIS) data could reveal the more spatial dimension of workers' (im)mobility. It could also visualize historic changes in workers' commuting patterns and job accessibility by mapping the workers' (affordable) housing locations, (entry-level) job locations, and mobility options in different periods.

Third, since this book is a first study of its kind, an international comparison based on secondary literature was difficult if not impossible. Where possible, I mentioned that general trends in the Netherlands diverged or converged with trends elsewhere. In order to compare and contrast more thoroughly the factors that enabled or disabled workers' mobility, and point out the unique or generic aspects of the Dutch case, further research is needed.

Fourth, in today's academic and policy debates on mobility poverty, the quantification of the phenomenon is key to understanding how many people are in this situation and what factors are most significant. *No Bus, No Bicycle, No Job* has made an important contribution to the historization of the phenomenon, also providing insight in past solutions and underlying political choices. But since this book took a qualitative approach, its aim was not to provide a thorough statistical understanding of how (im)mobility in twentieth century Netherlands affected job accessibility in particular, and social (in)equality in general. Current research and policies can, however, also gain from my qualitative research, as it helps draw a more holistic picture of the enabling and disabling factors in workers' mobility, sensitive to political ambiguity and contingency. I am therefore convinced that these and other limitations do not disqualify the findings of *No Bus, No Bicycle, No Job*, but rather serve as a preamble for future mobility justice-inspired research and offer interesting leads for historians of labor and mobility.

No Bus, No Bicycle, No Job moved beyond researching workers' mobility as mere movements from A to B or users of certain vehicles. It explored the multiplicity of practices, experiences, and meanings that shaped how workers got to work, and highlighted the constitutive role of power relations and labor market politics in workers' mobility in the Netherlands. This book has shown that the question of who leads and negotiates the production of mobility is the outcome of power relations and historic change, and also determines whose mobility is secured.

Although historically, employers have had an important role in reducing mobility barriers for those with few alternatives, their involvement often came with paternalistic intentions. I have detailed how in an industrial-capitalist system, the state, but mostly workers and employers solved mobility barriers. Today, multi-stakeholder coalitions that include these actors—like the postwar Miner Transportation Committee in Limburg—could help identify barriers and secure low-income workers' and jobseekers' mobility and job accessibility. In resolving today's predicaments and steering mobility transitions, local and supranational projects like the European Green Deal are key for reducing transport emissions while keeping citizens connected. Even though today we find ourselves in a post-industrial, neoliberal age, this study has provided ample evidence that, historically, workers and employers are potentially powerful agents of change, as long as their voices are heard in decision-making processes and their intentions are guided by ambitions of mobility justice rather than control.

Bibliography

Archival Collections

Fietsersbond Afdelingsarchieven: Archief en Onderzoek, Eindhoven University of Technology

- "Prijzen van fietsen (Kampioen diverse jaren)"

Gemeentearchief Vaals (GAV)

- Archive no. 2094, Archieven der Gemeente Vaals (1893) 1930-1981 (2000), Verkeer en Vervoer

International Institute for Social History, Amsterdam

- Archive no. ARCH01969, Archief ENFB

Katholiek Documentatie Centrum (KDC)

- Archive no. 370, Nederlands Katholiek Vakverbond (NKV), 1894-2001

Nationaal Archief (NA), The Hague

- Archive no. 2.16.83, Ministerie van Waterstaat: Bureau Autobussen, 1943-1950/ Commissie Vergunningen Personenvervoer, 1937-1959
- Archive no. 2.16.84, Ministerie van Waterstaat: Afdeling Vervoerwezen, 1941-1945
- Archive no. 3.17.17.04, Kamer van Koophandel en Fabrieken voor Rotterdam, 1922-2001
- Archive no. 2.15.62, Directoraat-Generaal voor de Arbeidsvoorziening van het Ministerie van Sociale Zaken en Werkgelegenheid, 1970-1990

NIOD Institute for War, Holocaust and Genocide Studies, Amsterdam

- Archive no. 039, Generalkommissariat für Finanz und Wirtschaft
- Archive no. 249-0881A, Dossier Vrijstellingen
- Archive no. 249-0198G, Dossier – Distributie – Rijwielen

Philips Company Archives (PCA), Eindhoven

- Archive no. 62.02, Personeel Philips, In en om de fabrieken: Uitgaan van de fabrieken
- Archive no. 642.5, Personeelsvervoer, Personeelsvervoer, Ned-NV VIPRE, and Snelbusdienst VIPRE
- Archive no. 642.11, Fietsen en fietsbanden
- Archive no. 728.14, Afdeling Vervoer en Garage

Regionaal Historisch Centrum Limburg (RHCL), Maastricht

- Archive no. 17.05H, Oranje-Nassau Mijnen, 1903-1978, Vervoer, 1903-1978
- Archive no. 17.05W, Oranje-Nassau Mijnen: Oorlogstijd en na-oorlogstijd, 1936-1966
- Archive no. EAN0065, Algemeene Bond van Christelijke Mijnwerkers in Nederland te Heerlen (1908-1940) en de Nederlandse Katholieke Mijnwerkersbond (NKMB) en voorgangers te Heerlen (1907-1975)

Stadsarchief Enschede (SE)

- Archive no. 61, C.B.I.P.E. te Enschede: Bedrijfsarchief

Tata Steel Central Archives (TSCA), IJmuiden

- Archive no. 4337, Personeelsvervoer (HFD)
- Archive no. 4338, Personeelsvervoer
- Archive no. 4349, Vervoer van en naar Beverwijk
- Archive no. 4342, Personeelsvervoer per fiets, bromfiets en motor (HFD)
- Archive no. 4344, Personeelsvervoer per autobus (HFD)
- Archive no. 4350, Vervoer van en naar Heemskerk, Uitgeest, Zaanstreek, Amsterdam

Online Collections

Delpher Newspaper Collection, 1618-2005, https://www.delpher.nl/nl/kranten
International Institute for Social History, Historisch Beeldarchief Migranten, http://www.iisg.nl/hbm/.
Krantenarchief de Staatsmijnen, http://krantenarchief.demijnen.nl/
Krantenkijker Gemeentearchief Schiedam, https://schiedam.courant.nu/
Mijnwerkerstijdschriften, http://www.shclimburg.nl/bibliotheek/mijnwerkerstijdschriften
Mobiliteits Museum, https://www.mobiel-erfgoed.nl/
Vijfeeuwenmigratie, https://vijfeeuwenmigratie.nl/

Newspaper and Journal Articles

"25 kilometer en niet harder." *Samen: Maandblad voor Directie en Personeel van Hoogovens en Nevenbedrijven* 29, no. 7 (July 1960): 132.
"Aankondiging Dienstregeling." *Limburgsch Dagblad*, 3 January 1931, 3.
"Aanleg van een spoorweg Sittard-Heerlen-Pruisische grens." *De Ingenieur* 29, no. 26 (1914): 511-512.
"Achterstand bij distributie van banden." *Algemeen Dagblad*, 16 October 1946, 1.
"Actie tegen verhoogde vervoerprijzen: Mijnwerkers boycotten autobusvervoer." *Leeuwarder Courant*, 17 May 1938, 10.
"Arbeiders wilden niet in een 'Rijksbus' rijden: Strijd om arbeidersvervoer in Borne." *Twentsch Dagblad Tubantia*, 17 December 1948, 2.
"Arbeidsbureau op wielen loopt gesmeerd." *Het Vrije Volk*, 14 January 1976, 17.
"Autobus-oorlog ten einde?" *De Telegraaf*, 17 October 1935, 3.
"Autobussen werden te duur." *Utrechts Volksblad: Sociaal-Democratisch Dagblad*, 17 May 1938, 1.
"Bedrijf en Maatschappij." *Mero-Meningen: Personeelsblad van N.J. Menko N.V. en N.V. Spinnerij Roombeek, Enschede* 8, no. 1 (1955): 3-4.
"Bedrijf en Maatschappij." *Mero-Meningen: Personeelsblad van N.J. Menko N.V. en N.V. Spinnerij Roombeek, Enschede* 8, no. 3 (1955): 15-16.
"Bijna 10.000 rijwielen; ruim 6000 gebreken." *Philips Koerier* 12, no. 7 (12 November 1955): 1-3.
"'Bijstandsvrouwen' met kinderen redden het niet." *Het Parool*, 23 October 1982, 7.
"Binnenland." *Het Vaderland*, 4 October 1925, 1.
"Bromfietsencontrole bracht vele gebreken aan het licht." *Philips Koerier* 15, no. 18 (14 February 1959): 2.
"Bus-vervoer mijnwerkers." *Limburger Koerier*, 22 March 1938, 6.
"Bus met mijnwerkers op vrachtauto." *Trouw*, 7 November 1957, 1.
"Busvervoer van mijnwerkers tot op de minuut geregeld." *Limburgsch Dagblad*, 16 December 1953, 5.
"Controle rijwielen en brommers." *Nieuws van de Staatsmijnen in Limburg* 13, no. 46 (11 December 1964): 7.
"Cursus voor bromfietsers." *Philips Koerier* 13, no. 3 (20 October 1956): 2.
"De hetze van de socialistische bond." *De Mijnwerker: Orgaan van de Ned. R.K. Mijnwerkersbond*, 4 June 1938, 178.
"De mijnwerker en z'n fiets." *Extra Uitgave van de Limburger Koerier*, 30 June 1932, 17.
"De veilige fiets." *Nieuws van de Staatsmijnen in Limburg* 6, no. 13 (29 March 1957): 8.
"De verhoogde tarieven voor mijnwerkersvervoer: In Echt en Roermond boycott der bussen." *Limburger Koerier*, 10 May 1938, 6.

"De vervoerkosten van en naar de mijnen." *Nieuwe Venlosche Courant*, 4 June 1938, 8.
"De werkliedentreinen naar Eindhoven." *Eindhovensch Dagblad*, 7 June 1930, 1.
"De wielrijder in het verkeer." *Nieuws van de Staatsmijnen in Limburg* 5, no. 11 (9 March 1956): 3.
"Distributie: Fietsbanden en fietsen." *De Tijd*, 14 February 1947, 2
"Diversen." *De Boerderij*, 23 June 1937, 33-36.
"Diversen." *De Boerderij*, 20 July 1938, 33-34.
"Een beroeps-fietsendief: Reeds een honderdtal karretjes gestolen." *Nieuwe Venlosche Courant*, 21 June 1930, 5.
"Een dringend beroep op alle weggebruikers." *Nieuws van de Staatsmijnen in Limburg* 4, no. 10 (7 May 1955): 2.
"Een eigen auto ... Waar doen ze het van?" *Spil en Spoel: Maandblad voor het Textielvak* 15, no. 10 (1965): 10-11.
"Eindelijk een échte Schriftelijke Verkeerscursus." *Mero-Meningen: Personeelsblad van N.J. Menko N.V. en N.V. Spinnerij Roombeek, Enschede* 9, no. 8 (1957): 8.
"Er wordt gebouwd." *Mero-Meningen: Personeelsblad van N.J. Menko N.V. en N.V. Spinnerij Roombeek, Enschede* 9, no. 10 (1957): 9.
"Extra treinen voor de Philips-arbeiders." *Eindhovensch Dagblad*, 10 July 1929, 1.
"Fiets contra autobus: Mijnwerkers tegen tarievenpolitiek." *Limburger Koerier*, 14 May 1938, 5.
"Fietsen naar het werk." *Het Parool*, 1 October 1975, 23.
"Fietsendieven hebben vrij spel." *De Telegraaf*, 12 February 1939, 3.
"Fouten tegen verkeersorde bevorderen onveiligheid." *Philips Koerier* 13, no. 16 (2 February 1957): 3.
"Geen richting aangeven: de meest gemaakte fout van wielrijders." *Philips Koerier* 13, no. 16 (2 February 1957): 1.
"Groen, geel, rood." *Samen: Maandblad voor Directie en Personeel van Hoogovens en Nevenbedrijven* 30, no. 1 (January 1961): 17.
"Heen en weer: Veel ongevallen tijdens spitsuren." *Philips Koerier* 15, no. 6 (8 November 1958): 2.
"Het belang van een goede stalling." *Het Volk*, 1 December 1941, 3.
"Het komt in de bus." *Samen: Maandblad voor Directie en Personeel van Hoogovens en Nevenbedrijven* 26, no. 7-8 (1957): 128-129.
"Het Personeelvervoer." *De Vrije Philips Koerier* 3, no. 8 (23 November 1946): 2.
"Het verkeer en U!" *Samen: Maandblad voor Directie en Personeel van Hoogovens en Nevenbedrijven* 28, no. 11 (November 1959 1959): 210-211.
"Het verkeer en U!" *Samen: Maandblad voor Directie en Personeel van Hoogovens en Nevenbedrijven* 29, no. 2 (February 1960): 30-31.
"Het verkeer en U!" *Samen: Maandblad voor Directie en Personeel van Hoogovens en Nevenbedrijven* 29, no. 5 (May 1960): 94-95.
"Het verkeer en U!" *Samen: Maandblad voor Directie en Personeel van Hoogovens en Nevenbedrijven* 29, no. 9 (September 1960): 170-171.
"Het vervoer der mijnwerkers." *Limburger Koerier*, 14 May 1938, 2.
"Het vervoervraagstuk der mijnwerkers." *De Mijnwerker: Orgaan van de Ned. R.K. Mijnwerkersbond*, 12 November 1938, 362.
"Honderd Turken in voormalig bejaardentehuis." *Schiedamsche Courant*, 13 January 1971, 3.
"In de bus." *Philips Koerier* 3, no. 44 (1946): 2.
"Is uw rijwiel wel in orde?" *Philips Koerier* 15, no. 2 (11 October 1958): 2.
"Jaarverslag over 1955 van de Veiligheidscommissie van N.J. Menko N.V. en N.V. Spinnerij Roombeek." *Mero-Meningen: Personeelsblad van N.J. Menko N.V. en N.V. Spinnerij Roombeek, Enschede* 8, no. 11 (1956): 9.
"'Kijk, dit is nou een onverharde weg ...' Turken worden ingewijd in de gevaren van het verkeer." *Het Vrije Volk*, 13 August 1965, 3.

"Kortweg." *De Kampioen* 86, no. 10 (October 1971): 580.
"Mijnwerkers boycotten autobussen: Felle twist in Zuid Limburg." *De Telegraaf*, 17 May 1938, 4.
"Na 20 jaar." *Spil en Spoel: Maandblad voor het Textielvak* 15, no. 10 (1965): 2.
"Nogmaals de autobus-kwestie." *De Mijnwerker: Orgaan van de Ned. R.K. Mijnwerkersbond*, 28 May 1938, 172.
"Officiële Medelingen: Sociale Rubriek." *Wilton-Fijenoord Nieuws*, 1 June 1953, 2.
"Ongevallen omlaag." *Mero-Meningen: Personeelsblad van N.J. Menko N.V. en N.V. Spinnerij Roombeek, Enschede* 8, no. 13 (1956): 3.
"Ongevallencijfer daalt." *Philips Koerier* 13, no. 26 (13 April 1957): 3.
"Onverwarmde mijnwerkerstreinen." *Algemeen Handelsblad*, 5 March 1919, 6.
"Onze zorgelijke rijwielpositie." *De Kampioen* 62, no. 3 (March 1947): 77-78.
"Ook in het verkeer naastenliefde?" *Spil en Spoel: Maandblad voor het Textielvak* 16, no. 5 (1965): 6-7.
"Openbaar streekvervoer per bus (deel 2)." *Consumentengids* September (1978): 422-424.
"Openstelling rijwielreparatiewerkplaats." *Staatsmijnen Nieuws* 16, no. 16 (21 July 1967): 10.
"Oplossing Veiligheidstekenpuzzle." *Samen: Maandblad voor Directie en Personeel van Hoogovens en Nevenbedrijven* 19, no. 2 (1950): 18-19.
"Pendelen." *Spil en Spoel: Maandblad voor het Textielvak* 13, no. 3 (1962): 4-5.
"Per fiets!" *Limburger Koerier*, 13 May 1938, 6.
"Philips' autobusdienst; 8 lijnen; 15 bussen." *Meierijsche Courant*, 18 February 1929, n.d.
"Philips eigen autobusdienst." *Eindhovens Dagblad*, 21 September 1928, 6.
"Philipspersoneel uit Gelderland." *Eindhovens Dagblad*, 28 February 1930, 2.
"Philipswagen Weert-Eindhoven." *Meierijsche Courant*, 16 February 1924, n.d.
"Protest van mijnwerkers tegen de nieuwe autobustarieven." *Limburgsch Dagblad*, 10 May 1938, 7.
"Regeling voor werklozen: Verhuiskosten vergoed bij nieuwe betrekking." *De Volkskrant*, 22 June 1977, 6.
"Rijwielbanden!" *Philips Koerier*, no. 1 (31 March 1943): 1.
"Rijwielbandenverstrekking." *De Vrije Philips Koerier* 2, no. 40 (13 July 1946): 4.
"Rijwielcontrole bij Wilton-Fijenoord in de Verkeersweek." *Wilton-Fijenoord Nieuws*, 1 November 1954, 33.
"Rijwielcontrole bracht ruim 6000 defecten aan het licht." *Nieuws van de Staatsmijnen in Limburg* 2, no. 25 (5 December 1953): 3.
"Rijwielcontrole op onze bedrijven." *Staatsmijnen Nieuws* 12, no. 47 (6 December 1963): 2.
"Rijwielcontrôle op onze bedrijven ontdekte ruim 4800 defecten." *Nieuws van de Staatsmijnen in Limburg* 1, no. 25 (5 December 1952): 2.
"Rijwielvergoedingen." *Philips Koerier*, 24 February 1943, 3.
"Rondom de autobus-kwestie." *De Mijnwerker: Orgaan van de Ned. R.K. Mijnwerkersbond*, 28 May 1938, 170.
"Sociale Luisterpost: Minder ongevallen." *Philips Koerier* 15, no. 18 (1959): 2.
"Susanneke werkt in Nederland." *De Vrije Philips Koerier* 3, no. 7 (16 November 1946): 2.
"Susteren: De mijnwerkers gaan per fiets." *Limburger Koerier*, 12 May 1938, 7.
"Tarieven mijnwerkersvervoer." *Limburger Koerier*, 10 May 1938, 5.
"Tarieven voor mijnwerkersbussen: Actie tegen de vastgestelde verhoogingen." *Limburger Koerier*, 29 March 1938, 6.
"Ter beteugeling van de verkeersonveiligheid: Beperkende voorschriften voor parkeren op complex Strijp te Eindhoven." *Philips Koerier* 14, no. 9 (30 November 1957): 8.
"Thorn: Autobusvervoer mijnwerkers." *Limburger Koerier*, 12 May 1938, 6.
"Twentsch Nieuws." *Twentsch Dagblad Tubantia*, 28 April 1916, 2.
"Vaarwel Hoogovenexpres: Herinneringen bij de laatste rit." *Samen: Maandblad voor Directie en Personeel van Hoogovens en Aangesloten Bedrijven* 26, no. 10 (1957): 163.

"Van het front der werklozen: Een mooi succes voor het Districtscomité voor Werklozenzorg." *De Fabrieksarbeider* 23, no. 50 (15 December 1934): 3.

"Veilig Verkeer." *Mero-Meningen: Personeelsblad van N.J. Menko N.V. en N.V. Spinnerij Roombeek, Enschede* 8, no. 6 (1955): 10.

"Veilig Verkeer." *Mero-Meningen: Personeelsblad van N.J. Menko N.V. en N.V. Spinnerij Roombeek, Enschede* 14, no. 5 (1960): 2.

"Veiligheid." *Wilton-Fijenoord Nieuws* 20 (1 October 1957): 17-18.

"Verkeerscursussen voor personeel in Eindhoven." *Philips Koerier* 15, no. 5 (1 November 1958).

"Verkeersles voor Turken." *Algemeen Dagblad,* 19 September 1964, 1.

"Verkeersles voor Turken." *Algemeen Dagblad,* 2 November 1972, 11.

"Verkeerssituate bij Staatsmijn Hendrik vraagt uiterste voorzichtigheid." *Nieuws van de Staatsmijnen in Limburg* 4, no. 10 (7 May 1955): 2.

"Verkeersstaking in Z.-Limburg: Weinig gewaardeerde ordening." *Algemeen Handelsblad,* 15 May 1938, 4.

"Vervoer van mijnwerkers: Bushouders willen prijsverhoging, mijnwerkers gaan per fiets." *Limburger Koerier,* 29 March 1938, 7.

"Voor- en Nadelen." *Spil en Spoel: Maandblad voor het Textielvak* 13, no. 3 (1962): 2.

"Wat leert een bezoek aan de mijnstreek? De productie moet en kan omhoog." *Amsterdamsch Dagblad,* 7 July 1945, 3.

"Wiecoda Rijwiel." *De Arbeider: Sociaal-Anarchistisch Weekblad,* 24 September 1932, 4.

"Zet het licht op groen." *Spil en Spoel: Maandblad voor het Textielvak* 9, no. 3 (1958): 36-37.

"Zij komen van verre." *Spil en Spoel: Maandblad voor het Textielvak* 13, no. 3 (1962): 1.

Arlman, Hugo, and Gerard Mulder. "Een warmkloppend hart voor de noden van de massa's." *Vrij Nederland,* no. 44 (5 November 1977): 4-5.

———. "Voor de mens als 'tool-using animal' is het gebruik van de auto heel natuurlijk." *Vrij Nederland,* no. 44 (5 November 1977): 26-27.

B. "Rijwielstalling Vlaardingerdijk." *Wilton-Fijenoord Nieuws,* 1 November 1954, 32.

Bakker Schut, Pieter. "Het tram-autobus-vraagstuk in verband met de stadsuitbreiding." *Wegen* 10, no. 5 (1934): 109-154.

Beckum, Paul van. "Ik pendel." *Verenigde Noordhollandse Dagbladen: Bijlage Kerstmis* (December 1965): 30-31.

Bergsma, Frits. "ROVER-hoofdman aan het woord: Sommigen zien ons als lastige pottekijkers." *Vogelvrije Fietser* 4, no. 4 (1979): 22-23.

Bijhouwer, Jan. "Bromfietsbespiegelingen." *Tijdschrift voor Volkshuisvesting en Stedebouw,* 12 December 1952, 172-173.

Boelsums, Anita, and Paulien Osse. "Bevrijdde het rijwiel de vrouw." *Vogelvrije Fietser* 2, no. 5 (1977): 14-15, 18-19.

Boer, Enne de. "Het onrecht van de sterkste: Sociale ongelijkheid in verkeer en vervoer." *Vogelvrije Fietser* 5, no. 3 (1980): 18-19, 28.

Buitelaar, J.M. "De mens en het monster." *Steenkool: Bedrijfstijdschrift van de Nederlandse Steenkolenmijnen* 6, no. 22 (23 November 1951): 506-507.

Crombach, Charles. "Gelijke vervoersmogelijkheden voor iederen." *Vogelvrije Fietser* 4, no. 2 (1979): 10-11.

Directie. "Ter introductie." *Wilton-Fijenoord Nieuws,* 1 October 1951, 2.

Dullemen, L.R. van. "Anderhalf miljoen brommers economisch nog te laag gewaardeerd." *De Kampioen* 78, no. 7 (July 1963): 408-409.

F.S. "Rijwielbezitters-medewerkers van W.F., let op!" *Wilton-Fijenoord Nieuws,* 1 October 1954, 24.

Groot, Gerard. "Massale volksverhuizing elke dag opnieuw in Nederland: Speciale busdiensten der bedrijven halen hun arbeiders van heinde en ver." *De Tijd,* 10 December 1955, 13.

H.S. "Immobiliteit maakt werkloosheid groter." *De Tijd,* 7 November 1972, 11.

Harthoorn, A.M. "Is het wenschelijk den locaalspoorweg Sittard-Herzogenrath tot hoofd-spoorweg te maken." *De Ingenieur* 31, no. 22 (1916): 413-415.
Kloeseman, W.F. "Wilton in het verkeer." *Wilton-Fijenoord Nieuws*, 1 October 1954, 21-22.
Krudop, E.R.S. "Wees een Heer in het verkeer." *Mero-Meningen: Personeelsblad van N.J. Menko N.V. en N.V. Spinnerij Roombeek, Enschede* 15, no. 8 (1962): 3-6.
Nieuwenhuis, Johan. "Het secundaire vervoer in crisistijd." *De Ingenieur* 51, no. 26 (1936): V27-V36.
Nooijen, H. van. "Een eigen huis." *De Tijd*, 30 June 1971, 5.
Nozeman, Ed. "Groeikernen: reisafstanden moesten kleiner worden maar het omgekeerde gebeurde." *ROM-Magazine*, no. 6 (June 1987): 8-12.
Ploeger, Jan. "Nederlandse fietsers neergelegd in cijfers." *Vogelvrije Fietser* 3, no. 4 (1978): 19-20.
S. "Veilig verkeer in en buiten het bedrijf." *Wilton-Fijenoord Nieuws*, 1 October 1948, 9.
Timmerman, J.B. "Verkeersveiligheid." *Wilton-Fijenoord Nieuws*, 1 October 1951, 2-3.
Verbond voor Veilig Verkeer. "Hangoren: Uw plaats op de weg." *De Wekker* 17, no. 2 (1963): 6.
———. "Houd rechts – Maak ruimte." *De Wekker* 15, no. 1 (1961): 7-8.
———. "Snel – sneller – snelst." *Spil en Spoel: Maandblad voor het Textielvak* 9, no. 11 (1959): 18-19.
———. "Snel – Sneller – Snelst." *De Wekker* 13, no. 2 (1959): 3.
———. "Snel ... sneller ... snelst." *Mero-Meningen* 11, no. 11 (1959): 12.
Wall, H. van de. "Over veiligheid gesproken .... Wijzer geworden." *Samen: Maandblad voor Directie en Personeel van Hoogovens en Nevenbedrijven* 27, no. 2 (February 1958): 31.
———. "Over veiligheid gesproken .... Alle-overschaduwend." *Samen: Maandblad voor Directie en Personeel van Hoogovens en Nevenbedrijven* 30, no. 4 (April 1961): 83.
———. "Over veiligheid gesproken .... Hoe hóórt het eigenlijk?" *Samen: Maandblad voor Directie en Personeel van Hoogovens en Nevenbedrijven* 35, no. 1 (January 1966): 17.
———. "Over veiligheid gesproken .... Leerzaam." *Samen: Maandblad voor Directie en Personeel van Hoogovens en Nevenbedrijven* 29, no. 2 (February 1960): 34.
———. "Over veiligheid gesproken .... Recht(s) is Recht(s)." *Samen: Maandblad voor Directie en Personeel van Hoogovens en Nevenbedrijven* 30, no. 9 (September 1961): 178.

Published Documentation of Government and Non-Governmental Organizations

Ambonezen in Nederland: Rapport van de Commissie ingesteld bij besluit van de Minister van Maatschappelijk Werk. The Hague: Staatsuitgeverij, 1959.
Wat moet u weten van het reizen met de VAV (Verenigd Arbeiders Vervoer) bussen. Heerlen: Gezamenlijke Steenkolenmijnen in Limburg, 1953.
Beuningen, Jacqueline van. "Automobilist het snelst op plaats bestemming." Centraal Bureau voor de Statistiek (Den Haag: CBS, 2013). Accessed June 17, 2021. https://www.cbs.nl/nl-nl/achtergrond/2013/28/automobilist-het-snelst-op-plaats-bestemming.
Bijl, Rob van der, and Hugo van der Steenhoven. *Gesprekken over gebrekkige mobiliteit: Vervoersarmoede in de grote stad ontrafelen.* Amsterdam/Utrecht: Favas/HugoCycling, 2019.
Boer, Enne de. "Mobiel en niet-mobiel: Een verkenning van de sociale betekenis van ons vervoer." In *Stedelijk verkeer en vervoer langs nieuwe banen? Preadviezen voor het symposium, 5 oktober 1976*, edited by J. Overeem, 47-78. The Hague: Stichting Toekomstbeeld der Techniek, 1976.
———. "Vervoersongelijkheid, relevantie en onderzoekbaarheid: Een verkenning van de internationale literatuur." The Hague: Projectbureau Integrale Verkeers en Vervoersstudies, Ministerie van Verkeer en Waterstaat, 1980.

Bosch, E.F. "De sociale problematiek van Tholen en Sint-Philipsland, Deel 1: Het Forensisme." Middelburg: Stichting Zeeland voor Maatschappelijk en Cultureel Werk, 1957.

Centraal Bureau voor de Statistiek. *Statisch Zakboek 1935*. The Hague: Martinus Nijhoff, 1935.

———. *Statisch Zakboek 1938*. The Hague: Martinus Nijhoff, 1938.

———. *12e Volkstelling annex woningtelling, 31 Mei 1947, Serie B: Voornaamste cijfers per gemeente/ Deel 6: Beroepsbevolking naar woon-, werk- en geboortegemeente*. Utrecht: De Haan, 1952.

———. *13e Algemene volkstelling, 31 mei 1960, Deel 11: Buiten de woongemeente werkenden*. Hilversum: De Haan, 1965.

———. *De mobiliteit van de Nederlandse bevolking in 1978*. The Hague: Staatsuitgeverij, 1981.

———. *14e Algemene volkstelling annex woningtelling, 28 februari 1971, Deel 7B: Woon-werkverkeer : voornaamste uitkomsten per gemeente*. The Hague: Staatsuitgeverij, 1982.

ENFB, Fietsersbond. *Werken met de fiets*. The Hague: Hoofdafdeling Infrastructuur en Verkeer, Ministerie van Verkeer en Waterstaat, 1992.

Eys, Henriette van. *Stedelijke Bereikbaarheid: Een voorstel tot kwaliteitsverbetering en kostenbesparing in het verkeers- en vervoerssysteem te Rotterdam*. Rotterdam: Stad en Landschap, 1983.

———. *Rotterdam, een stad voor iedereen? Een onderzoek naar een rechtvaardiger verdeling van ruimte en voorzieningen voor alle stadsbewoners*. Rotterdam: Dienst Stadsontwikkeling, 1981.

———. *Emancipatoire aandachtspunten in stedebouwkundige plannen*. Rotterdam: Gemeente Rotterdam, 1986.

Haas, Matthijs de, and Marije Hamersma. "Fietsfeiten: nieuwe inzichten." The Hague: Kennisinstituut voor Mobiliteitsbeleid, 2020.

Hoorn, A.I.J.M. van der, and M.R. Mulder. "Een gedisaggregeerde registratie van de ontwikkeling van rijbewijs- en autobezit." The Hague: Dienst Verkeerskunde, Rijkswaterstaat, 1981.

Jorritsma, Peter, Jaco Berveling, Mathijs de Haas, Peter Bakker, and Lucas Harms. *Mobiliteitsarmoede: Vaag begrip of concreet probleem?*. The Hague: Kennisinstituut voor Mobiliteitsbeleid, 2018.

Katteler, H., W.F. de Heer, and J. Kropman. *Het gebruik van de fiets in Nederland: Een aanvullend onderzoek in de provincies Groningen, Friesland, Zuid-Holland, Zeeland, Noord-Brabant en Limburg*. Nijmegen: ITS, 1978.

Lambooy, J.G., and J.H. Mak. "De pendelaar in woon-, reis-, en werkomgeving." IJmuiden: Koninklijke Nederlandsche Hoogovens en Staalfabrieken, n.d.

Marle, Th. M. B. van. *Het binnenlands vervoer na de bevrijding: 1944/45-1952*. The Hague: Nederlands Verkeersinstituut, 1953.

Meilof, Roel. *Bijlage: Wegwijzer woon-werkverkeer*. The Hague: Commissie Ontwikkeling Bedrijven van de Sociaal-Economische Raad, 1991.

Ministerie van Verkeer en Waterstaat. *Structuurschema Verkeer en Vervoer*. The Hague: Tweede Kamer der Staten-Generaal, 1976-1977.

———. *Meerjarenplan voor het personenvervoer 1976-1980*. The Hague: Staatsuitgeverij, 1975.

———. *Structuurschema Verkeer en Vervoer: Deel C, Adviezen van de Raad van de Waterstaat en van de Raad van Advies voor de Ruimtelijke Ordening*. The Hague: Staatsuitgeverij, 1978-1979.

———. "Werkgevers stimuleren fietsgebruik medewerkers." Rijksoverheid. Accessed June 18, 2021. https://www.rijksoverheid.nl/onderwerpen/fiets/werkgevers-stimuleren-fietsgebruik-medewerkers.

Molnár-in 't Veld, Hermine. "De groei van het Nederlandse personenautopark." Centraal Bureau voor de Statistiek. (Den Haag: CBS, 2019). Accessed June 21, 2021. https://www.cbs.nl/nl-nl/longread/statistische-trends/2019/de-groei-van-het-nederlandse-personenautopark/3-resultaten.

MuConsult, *Mobiliteit van allochtonen: Verslag van een voorstudie*. Utrecht: MuConsult, 1994.

Nederlandse Stichting voor Statistiek. *De buitenlandse arbeider in Nederland*. The Hague: Staatsuitgeverij, 1971.

Roemer, Emile. *Geen Tweederangsburgers: Aanbevelingen om misstanden bij arbeidsmigranten in Nederland tegen te gaan.* The Hague: Aanjaagteam Bescherming Arbeidsmigranten, Rijksoverheid, 2020.

ROVER. *Gemiste Bussen: Een nota van de Werkgroep Streekvervoer van de Vereniging Reizigers Openbaar Vervoer (ROVER) met als onderwerp: Het voorzieningenniveau van het streekvervoer in Nederland.* Haarlem: Stichting ROVER, 1982.

Schouten, N.C. "Pendelarbeiders en industrievestiging op Tholen." Middelburg: Stichting Zeeland voor Maatschappelijk en Cultureel Werk, 1961.

Schreven, Luuk. *Documentatierapport Volkstelling 1971.* The Hague: Centraal Bureau voor de Statistiek, 2004.

Stichting Werkgroep 2Duizend. *Brochure: Werken met de fiets: Evaluatie van de ontvangst door de doelgroep, eindrapport.* Amersfoort, Stichting Werkgroep 2Duizend, 1993.

Tuinstra, Fons. *Woon-werkverkeer in schonere banen.* Amsterdam: FNV Centrum Ondernemingsraden, 1990.

Verbond voor Veilig Verkeer. *De Onderneming en de Verkeersveiligheid.* Utrecht: Verbond voor Veilig Verkeer, 1958.

Verkeersonderneming, De. "Werkgeversaanpak." Accessed May 7, 2021. https://www.verkeersonderneming.nl/r10/.

Wetenschappelijke Raad voor het Regeringsbeleid. *Rapporten aan de Regering: Vernieuwingen in het arbeidsbestel.* The Hague: Staatsuitgeverij, 1981.

Scholarly Publications

Albert de la Bruhèze, Adri A. "Enschede: An Experiment in Cycling" In *Cycling Cities: The European Experience.* Edited by Ruth Oldenziel, Martin Emanuel, Adri Albert de la Bruhèze, and Frank Veraart, 41-51. Eindhoven: Foundation for the History of Technology, 2016.

Albert de la Bruhèze, Adri, and Frank Veraart. *Fietsverkeer in praktijk en beleid in de twintigste eeuw.* The Hague: Foundation for the History of Technology, 1999.

Anderson, Benedict. *Imagined Communities: Reflections of the Origin and Spread of Nationalism.* London: Verso, 1983.

Arora, Saurabh, Barbara van Dyck, Divya Sharma, and Andy Stirling. "Control, Care, and Conviviality in the Politics of Technology for Sustainability." *Sustainability: Science, Practice and Policy* 16, no. 1 (2020): 247-262.

Atzema, Oedzge, and Egbert Wever. *De Nederlandse industrie: Vernieuwing, verwevenheid en spreiding.* Assen: Van Gorcum, 1999.

Baker, Robert. "Socialism in the Nord, 1880–1914. A Regional View of the French Socialist Movement." *The International Review of Social History* 12, no. 3 (1967): 357-389.

Bakker Schut, Frits. *Industrie en woningbouw: Een technisch-economische beschouwing over bemoeiingen van de industrie met arbeiderswoningbouw.* Assen: Van Gorcum & Comp., 1933.

Barlowe-Kayes, Jamie. "Reading against the Grain: The Powers and Limits of Feminist Criticism of American Narratives." *The Journal of Narrative Technique* 19, no. 1 (1989): 130-140.

Bastiaanssen, Jeroen, Daniel Johnson, and Karen Lucas. "Does Transport Help People to Gain Employment? A Systemic Review and Meta-Analysis of the Emperical Evidence." *Transport Reviews* 40, no. 5 (2020): 607-628.

Bastiaanssen, Jeroen, Karel Martens, and Gert-Jan Polhuijs. "'Geen rijbewijs, geen fiets, geen ov-aansluiting, geen baan': Vervoersarmoede in Rotterdam-Zuid." *Verkeerskunde* 63, no. 5 (2013): 44-50.

Berkers, Eric. "Fietscultuur en -beleid in Maastricht en Parkstad in historisch perspectief." Eindhoven: Foundation for the History of Technology, 2017.

Berkers, Eric, and Ruth Oldenziel. *Cycling Cities: The Arnhem and Nijmegen Experience.* Eindhoven: Foundation for the History of Technology, 2017.

Berkers, Eric, Frank Schipper, Patrick Bek, and Ruth Oldenziel. *Cycling Cities: The Rotterdam Experience.* Eindhoven: Foundation for the History of Technology, 2019.

Bervoets, Elisabeth. "Modernisering van de woningbouw 1890-1970: een bijzonder patroon van technische vernieuwing." In *Techniek in Nederland in de twintigste eeuw. Deel 6. Stad, bouw, industriële productie*, edited by Johan Schot, Harry Lintsen, Arie Rip, and Adri Albert de la Bruhèze, 110-117. Eindhoven/Zutphen: Foundation for the History of Technology/Walburgpers, 2003.

Blok, Els. *Loonarbeid van vrouwen in Nederland 1945-1955.* Nijmegen: SUN, 1978.

Blok, Johan, Cor Campagne, and Sjef Janssen. *Trams in Midden- en Zuid-Limburg 1888-1950.* Heerlen: Schuyt & Co. Uitgevers, 1998.

Boer, Enne de. *Transport Sociology: Social Aspects of Transport Planning.* Oxford: Pergamon, 1986.

Bond, Winstan. "The Flawed Economics and Morality of the American Uniform Five-cent Fare." In *Suburbanizing the Masses: Public Transport and Urban Development in Historical Perspective*, edited by Colin Divall and Winstan Bond, 49-78. London/New York: Routledge, 2003.

Bonham, Jennifer. "Transport: Disciplining the Body That Travels." *The Sociological Review* 54, no. 1 (2006): 55-74.

Bonham, Jennifer, and Peter Cox. "The Disruptive Traveller? A Foucauldian Analysis of Cycleways." *Road and Transport Research* 19, no. 2 (2010): 42-53.

Borg, Kurt. "Conducting Critique: Reconsidering Foucault's Engagement with the Question of the Subject." *Symposia Melitensia*, no. 11 (2015): 1-15.

Bouwens, Bram, Joost Dankers, Yvonne van Mil, Reinout Rutte, Katie Sluyterman, and Jaap Verheul. *Door staal gedreven: Van Hoogovens tot Tata Steel, 1918-2018.* Bussum: Uitgeverij Thoth, 2018.

Bouwens, Sophie. *Over de streep: Grensarbeid vanuit Zuid-Limburg naar Duitsland, 1959-2001.* Maaslandse Monografieën. Vol. 71. Hilversum: Verloren, 2011.

Breman, Jan. "Industrial Labour in Post-Colonial India II: Employment in the Informal-Sectory Economy." *The International Review of Social History* 44, no. 3 (1999): 451-483.

Brömmelstroet, Marco te. "Een andere kijk op fietsen: Over stofzuigers, ganzenlogica en spreeuwenzwermen." Cycling Academics Blog, 2014. Accessed August 30, 2021. http://cyclingacademics.blogspot.com/2014/06/een-andere-kijk-op-fietsen-over.html.

———. *Mobility Language Matters.* Amsterdam: De Correspondent Uitgevers, 2020.

Buiter, Hans. "Werken aan sanitaire en bereikbare steden, 1880-1914." In *Techniek in Nederland in de twintigste eeuw. Deel 6: Stad, bouw, industriële productie*, edited by Johan Schot, Harry Lintsen, Arie Rip and Adri Albert de la Bruhèze, 25-49. Eindhoven/Zutphen: Foundation for the History of Technology/Walburgpers, 2003.

Buren, Frank van. "Kolenwinning onder Duitse druk: productie- en arbeidsverhoudingen in de Limburgse mijnen tijdens de bezetting 1940-1944." *Studies over de sociaal-economische geschiedenis van Limburg* XLIV (2004): 135-164.

Capuzzo, Paolo. "Transport and (sub)urban development: Between politics and technology: transport as a factor of mass suburbanization in Europe, 1890-1939." In *Suburbanizing the Masses: Public Transport and Urban Development in Historical Perspective* edited by Colin Divall and Winstan Bond, 23-48. London/New York: Routledge, 2003.

Caubo, Harry. "Mijn vader, de mijnwerker." Heerlen Vertelt. Accessed June 16, 2021. http://www.heerlenvertelt.nl/2016/06/mijn-vader-de-mijnwerker/.

Chin, Rita. *The Guest Worker Question in Postwar Germany.* Cambridge: Cambridge University Press, 2009.

Constantine, Simon. "Migrant Labour in the German Countryside: Agency and Protest, 1890–1923." *Labor History* 47, no. 3 (2006): 319-341.

Cowie, Jefferson, and Joseph Heathcott. "The Meanings of Deindustrialization." In *Beyond the Ruins: The Meanings of Deindustrialization*, edited by Jefferson Cowie and Joseph Heathcott, 1-15. Ithaca, NY: ILR Press, 2003.

Cresswell, Tim. "Towards a Politics of Mobility." *Environment and Planning D: Society and Space* 28, no. 1 (2010): 17-31.

Daunton, Martin. "Miners' Houses: South Wales and the Great Nortern Coalfield, 1880-1914." *The International Review of Social History* 25, no. 2 (1980): 143-175.

Davies, Reg. *Public Passenger Transport in Inter-war Britain: The Southern Railway's Response to Bus Competition, 1923-39.* PhD Dissertation, University of York, 2014.

Deben, Leon, and Ineke Teijmant. "Wonen is meer dan een dak boven je hoofd." In *Vertrouwde patronen, nieuwe dromen: Nederland naar een modern industriële samenleving 1948-1973*, edited by Hélène Vossen, Marjan Schwegman, and Peter Wester, 134-141. IJsselstein: VGN, 1992.

De Block, Greet. "Urbanizing the Countryside: Rails, Workers and Commuting in South-West Flanders, Belgium, 1830-1930." In *Cultural Histories of Sociabilities, Spaces and Mobilities*, edited by Colin Divall, 53-66. London: Routledge, 2015.

De Block, Greet, David de Kool, and Bruno de Meulder. "Paradise Regained?". In *Builders and Planners: A History of Land-use and Infrastructure Planning in the Netherlands*, edited by Jos Arts, Ruud Filarski, Hans Jeekel and Bert Toussaint, 17-68. The Hague: Eburon Academic Publishers, 2016.

Dekker, Henk-Jan. "An Accident of History? How Mopeds Boosted Dutch Cycling Infrastructure (1950–1970)." *The Journal of Transport History* (April 2021): 1-24.

———. *Cycling Pathways: The Politics and Governance of Dutch Cycling Infrastructure, 1920-2020.* Amsterdam: AUP, 2021.

Delfgaauw, Gerardus. *De tendenzen tot decentralisatie in de vestiging der nijverheid.* Serie voor Stedebouw VII. Amsterdam: Nederlandsch Instituut voor Volkshuisvesting en Stedebouw, 1932.

Dinius, Oliver, and Angela Vergara. *Company Towns in the Americas: Landscape, Power, and Working-Class Communities.* Athens, GA: University of Georgia Press, 2011.

Dinubila, Elena. "Fight for Meaning: Representations and Work Experiences in a Greenfield Automotive Plant." *Labor History* 61, no. 1 (2020): 60-73.

Divall, Colin, and Winstan Bond. *Suburbanizing the Masses: Public Transport and Urban Development in Historical Perspective.* London/New York: Routledge, 2003.

Divall, Colin, and Barbara Schmucki. "Introduction." In *Suburbanizing the Masses: Public Transport and Urban Development in Historical Perspective* edited by Colin Divall and Winstan Bond, 1-22. London/New York: Routledge, 2003.

Edgerton, David. *Shock of the Old: Technology and Global History Since 1900.* Oxford: Oxford University Press, 2006.

Ekman, Tomas. "Vision in Solid Form: A Comparison Between Two Solutions to the Traffic Probelm in Stockholm, 1941 and 1992." In *Suburbanizing the Masses: Public Transport and Urban Development in Historical Perspective*, edited by Colin Divall and Winstan Bond, 171-186. London/New York: Routledge, 2003.

Elteren, Melchior van. *Staal en Arbeid: Een sociaal-historische studie naar industriële accomodatieprocessen onder arbeiders bij het desbetreffend bedrijfsbeleid bij Hoogovens IJmuiden, 1924-1966.* Leiden: E.J. Brill, 1986.

Emanuel, Martin. "Copenhagen: Branding the Cycling City." In *Cycling Cities: The European Experience*, edited by Ruth Oldenziel, Martin Emanuel, Adri Albert de la Bruhèze, and Frank Veraart, 77-87. Eindhoven: Foundation for the History of Technology, 2016.

———. "Stockholm: Where Public Transit Eclipses Cycling." In *Cycling Cities: The European Experience*, edited by Ruth Oldenziel, Martin Emanuel, Adri Albert de la Bruhèze, Frank Veraart, 149-159. Eindhoven: Foundation for the History of Technology, 2016.

Ewalds, Doreen, Ger Moritz, and Michel Sijstermans. *Bromfietsen in Nederland*. The Hague/ Heerlen: Centraal Bureau voor Statistiek, 2013.

Fari, Simone, and Massimo Moraglio. "Future Mobilities: A Challenge for Economic and Business Historians." In *43rd Annual Economic and Business History Society Conference*. Jyväskylä: University of Jyväskylä, 2018.

Filarski, Ruud. "De coördinatiecrisis." In *Van transport naar mobiliteit: De mobiliteitsexplosie (1895-2005)*, edited by Gijs Mom and Ruud Filarski, 203-235. Zutphen: Walburgpers, 2008.

———. "The Emergence of the Bus Industry: Dutch Transport Policy During the Interwar Years." *Transfers* 1, no. 2 (2011): 61-82.

First, Ruth. "The Bus Boycott." *Africa South* (July–September 1957): 55-64.

Flonneau, Mathieu. "Collective Taxis in 1930s Paris: A Contribution to an Archeology of 'Uberization'." *The Journal of Transport History* 39, no. 1 (2018): 12-24.

Foucault, Michel. *Discipline and Punish: The Birth of the Prison*. London: Penguin/Random House, 2020 [1977].

———. *History of Sexuality*. Vol. 1, London: Penguin Books, 1978.

———. *Power/Knowledge: Selected Interviews and Other Writings, 1972-1977*. New York: Pantheon Books, 1980.

Fuchs, J.M. *Onderweg: Een eeuw personen- en goederenvervoer in Nederland*. Amsterdam: KNVTO, 1981.

Furness, Zack. "Biketivism and Technology: Historical Reflections and Appropriations." *Social Epistemology* 19, no. 4 (2005): 401-417.

Gillissen, J.M.A.G. *De invloed van de Staatsmijnen in Limburg op de ontwikkeling van het locale personenvervoer en op de woningbouw in de mijnstreek tot negentien honderd en veertig*. PhD Dissertation, Sociaal Historisch Centrum Limburg, 1962.

Gitelman, Howard M. "Welfare Capitalism Reconsidered." *Labor History* 33, no. 1 (1992): 5-31.

Gommers, Michael, and Piet Bovy. *Vervoermiddelgebruik bij woonforensen uit kleine kernen*. Delft: Delftse Universitaire Pers, 1988.

Groot, Gertjan de. *Fabricage van verschillen: Mannenwerk, vrouwenwerk in de Nederlandse industrie (1850-1940)*. Amsterdam: Aksant, 2001.

Gunn, Simon. "People and the Car: The Expansion of Automobility in Urban Britain, c.1955–70." *Social History* 38, no. 2 (2013): 220-237.

Hägerstrand, Torsten. *The Impact of Transport on the Quality of Life*. Lund: Lunds Universitets Kulturgeografiska Institution, 1974.

Have, Wichtert ten. *1940: Verwarring en aanpassing*. Series *Leven in bezet Nederland*. Edited by Erik Somers and René Kok. Houten/Amsterdam: Spectrum/NIOD, 2015.

Herod, Andrew. "Social Engineering through Spatial Engineering: Company Towns and the Geographical Imagination." In *Company Towns in the Americas: Landscape, Power, and Working-Class Communities*, edited by Oliver Dinius and Angela Vergara, 21-44. Athens, GA: University of Georgia Press, 2011.

Heuvel, Mark van den. "Verschuivende contexten van vrije tijd en vermaak." In *Vertrouwde patronen, nieuwe dromen: Nederland naar een modern industriële samenleving 1948-1973*, edited by Hélène Vossen, Marjan Schwegman and Peter Wester, 97-104. IJsselstein: VGN, 1992.

Hillman, Mayer. "Social Goals for Transport Policy." Paper presented at the Transport for Society Conference, London, 1975.
Hoel, L.A., E.D. Perle, K.J. Kansky, A.A. Kuehn, E.S. Roszner, and H.P. Nesbitt. *Latent Demand for Urban Transportation.* Pittsburg, PA: Carnegie Mellon University, 1968.
Hoeven, A. ter. "De positie van de pendelarbeider op de locale arbeidsmarkt." *Sociologische Gids* 6, no. 1 (1959): 17-30.
Hoeven, Pieter ter. *De Havenarbeiders van Amsterdam en Rotterdam: Een sociologische analyse van een arbeidsmarkt.* Rotterdam: Stenfert Kroese, 1963.
Hulten, Michel van. *'Gratis' openbaar vervoer.* Deventer: Kluwer, 1972.
Huré, Maxime. "Lyon: The Bicycle Goes Corporate." In *Cycling Cities: The European Experience,* edited by Ruth Oldenziel, Martin Emanuel, Adri Albert de la Bruhèze, and Frank Veraart, 173-183. Eindhoven: Foundation for the History of Technology, 2016.
Hyde, Charles. "Planning a Transportation System for Metropolitan Detroit in the Age of the Automobile: The Triumph of the Expressway." *Michigan Historical Review* 32, no. 1 (2006): 59-95.

Illich, Ivan. *Deschooling Society.* San Francisco: Harper & Row, 1971.
———. *Tools for Conviviality.* San Francisco: Harper & Row, 1973.

Jager, H. de. "Personenvervoer bij de Hoogovens." *Rail Magazine,* no. 174 (2000): 20-23.
Jeekel, Hans. *Inclusive Transport: Fighting Involuntary Transport Disadvantages.* Cambridge, MA: Elsevier, 2018.
Jeekel, Hans, and Karel Martens. "Equity in Transport: Learning from the Policy Domains of Housing, Health Care and Education." *European Transport Research Review* 9, no. 4 (2017): 1-13.
Jong, Harmen Jan de. *De Nederlandse industrie 1913-1965: Een vergelijkende analyse op basis van productiestatistieken.* PhD Dissertation, Rijksuniversiteit Groningen, 1999.
Jordan, Pete. *De Fietsrepubliek.* Amsterdam: Uitgeverij Podium, 2013.
Judt, Tony. *Postwar: A History of Europe Since 1945.* London: Random House, 2010 [2005].

Kain, John. "Housing Segregation, Negro Employment, and Metropolitan Decentralization." *Quarterly Journal of Economics* 82, no. 2 (1968): 175-197.
Kasper, Hans, Henri Faun, Janneke Gardeniers, Stéphanie Knoors, Leen Roels, Ad Knotter, and Willibrord Rutten. *Na de mijnsluiting in Zuid-Limburg: 35 jaar herstructurering en reconversie 1965-2000 en een doorkijk naar 2010.* Maastricht: Stichting Behoud Mijnhistorie, 2013.
Kenyon, Susan, Glenn Lyons, and Jack Rafferty. "Transport and Social Exclusion: Investigating the Possibility of Inclusion through Virtual Mobility." *Journal of Transport Geography* 10, no. 3 (2003): 207-219.
Kerr, Ian. "On the Move: Circulating Labor in Pre-Colonial, Colonial, and Post-Colonial India." *The International Review of Social History* 51, Supplement 14 (2006): 85-109.
Klemann, Hein. *De Nederlandse economische contacten met Nazi-Duitsland en de Nederlandse economie tijdens de Tweede Wereldoorlog: Een bibliografie.* Amsterdam: NEHA, 1995.
Klerk, Len de, and Ries van der Wouden. *Ruimtelijke ordening: Geschiedenis van de stedelijke en regionale planning in Nederland, 1200-nu.* Rotterdam: nai010, 2021.
Kuijs, Gerard. *De vrees voor wat niet kwam: Nieuwe arbeidsverhoudingen in Nederland 1935-1945, aan het voorbeeld van de Twentse textielindustrie.* Amsterdam: Vossiuspers UvA, 2010.
Kuttler, Tobias, and Massimo Moraglio. *Re-Thinking Mobility Poverty: Understanding Users' Geographies, Backgrounds and Aptitudes.* London/New York: Routledge, 2020.

Langendonck, Jef van. *Passende arbeid: Een verkenning in de wetgeving van een aantal Europese landen.* Sociale Zekerheidsreeks. Deventer: Kluwer, 1980.

Langeweg, Serge. "De geographische spreiding van de mijnwerkersbevolking." In *Mijnwerkers in Limburg: Een Sociale Geschiedenis*, edited by Ad Knotter, Serge Langeweg, Erik Nijhof, Jan Peet, and Willibrord Rutten, 100-138. Nijmegen: Van Tilt, 2012.

———. *Mijnbouw en arbeidsmarkt in Nederlands-Limburg: Herkomst, werving, mobiliteit en binding van mijnwerkers tussen 1900 en 1965*. Hilversum: Verloren, 2011.

Lavenir, Cathérine. "Scarcity, Poverty, Exclusion: Negative Associations of Bicycle's Uses and Cultural History in France." In *Cycling and Recycling: Histories of Sustainable Practices*, edited by Ruth Oldenziel and Helmuth Trischler, 58-71. New York/Oxford: Berghahn Books, 2016.

Leeuwen, Bas van, Robin Philips, and Erik Buyst. *An Economic History of Regional Industrialization.* London: Routledge, 2021.

Leerink, Jan. *De verkeers-veiligheid op den weg: Een juridische, sociologische en verkeerstechnische studie (proefschrift)*. Alphen aan den Rijn: N. Samsom, 1938.

Linden, Marcel van der. *Workers of the World: Essays Toward a Global Labor History.* Leiden/ Boston: Brill, 2011.

Linden, Marcel van der. "Work Incentives and Forms of Supervision." In *Handbook Global History of Work*, edited by Karin Hofmeester and Marcel van der Linden, 469-489. Berlin/Boston: De Gruyter Oldenourg, 2018.

Lucas, Karen. "The Role of Transport in the Social Exclusion of Low Income Populations in South Africa." Paper presented at the *World Conference on Transport Research Society*, Lisbon, Portugal, July 11-15, 2010. https://www.wctrs-society.com/wp-content/uploads/abstracts/lisbon/general/01390.pdf.

———. *Running On Empty: Transport, Social Exclusion and Environmental Justice.* Bristol: The Policy Press, 2004.

———. "Transport and Social Exclusion: Where Are We Now?" *Transport Policy* 20, March (2012): 105-113.

Lüdtke, Alf, and William Templer. "Polymorphous Synchrony: German Industrial Workers and the Politics of Everyday Life." *The International Review of Social History* 38, Supplement 1 (1993): 39-84.

Maier, Charles S. "Between Taylorism and Technocracy: European Ideologies and the Vision of Industrial Productivity in the 1920s." *Journal of Contemporary History* 5, no. 2 (1970): 27-61.

Manuhutu, Wim, and Henk Smeets. *Tijdelijk verblijf. De opvang van Molukkers in Nederland, 1951.* Amsterdam: De Bataafsche Leeuw, 1991.

Marcuse, Herbert. *One-Dimensional Man.* London/ New York: Routledge, 2002 [1964].

Martens, Karel. "Role of the Bicycle in the Limitation of Transport Poverty in the Netherlands." *Journal of the Transportation Research Board* 2387, no. 1 (2013): 20-25.

———. *Transport Justice: Designing Fair Transportation Systems.* New York/London: Routledge, 2016.

Martens, Karel, Marnix ten Holder, and Jurriën Thijssen. "Vervoersarmoede Bestaat." *Verkeerskunde* 2, no. 11 (2011): 34-38.

Messing, Frans. *Geschiedenis van de mijnsluiting in Limburg: Noodzaak en lotgevallen van een regionale herstructurering 1955-1975*. Leiden: Martinus Nijhoff, 1988.

Mohun, Arwen. "Labor and Technology." In *A Companion to American Technology* edited by Carroll Pursell, 212-230. Malden, MA: Blackwell, 2008.

Mom, Gijs. "Clashes of Cultures: Road vs. Rail in the North Atlantic World during the Interwar Coordination Crisis." In *The Organization of Transport: A History of Users, Industry, and Public History*, edited by Massimo Moraglio and Christopher Kopper, 18-37. New York: Routledge, 2015.

———. "Roads Without Rails: European Highway Network-Building and the Desire for Long-Range Motorized Mobility." *Technology and Culture* 46, no. 4 (2005): 745-772.
———. "What Kind of Transport History Did We Get? Half a Century of JTH and the Future of the Field." *The Journal of Transport History* 24, no. 2 (2003): 121-138.
———. "The Emancipation of the Urban View: Dutch Spatial Planning in an International Context (1920-1950)." In *Builders and Planners: A History of Land-use and Infrastructure Planning in the Netherlands*, edited by Jos Arts, Ruud Filarski, Hans Jeekel and Bert Toussaint, 73-139. The Hague: Eburon Academic Publishers, 2016.
Mom, Gijs, Colin Divall, and Peter Lyth. "Towards a Paradigm Shift? A Decade of Transport and Mobility History." In *Mobility in History: The State of the Art in the History of Transport, Traffic and Mobility*, edited by Gijs Mom, Gordon Pirie, and Laurent Tissot, 13-40. Neuchâtel, Editions Alphil, 2009.
Mom, Gijs, and Ruud Filarski. *Van Transport naar Mobiliteit: De Mobiliteitsexplosie 1895-2005*. Zutphen: Walburgpers, 2008.
Mom, Gijs, Johan Schot, and Peter-Eloy Staal. "Werken aan mobiliteit: de inburgering van de auto." In *Techniek in Nederland in de twintigste eeuw: Transport, communicatie*, edited by Harry Lintsen and Johan Schot, 45-73. Zutphen: Walburgpers, 2002.
Moraglio, Massimo. "Rails in the Car Kingdom: Competition, Changes and Continuity in Urban Mobility – The Case of Turin, 1914-1973." In *From Rail to Road and Back Again? A Century of Transport Competition and Interdependency*, edited by Ralf Roth and Colin Divall, 355-372. London/New York: Routledge, 2015.
Morgan, Njogu. *Cycling Cities: The Johannesburg Experience*. Eindhoven: Foundation for the History of Technology, 2019.
Morris, Kate. "Research into travel horizons and its subsequent influence on accessibility planning and demand responsive transport strategies in Greater Manchester." Paper presented at the *European Transport Conference*. Strasbourg, France, 2006.

Nacar, Can. ""Our Lives Were Not as Valuable as an Animal': Workers in State-Run Industries in World-War-II Turkey." *The International Review of Social History* 54, Supplement (2009): 143-166.
Nijhof, Erik. *'Gezien de dreigende onrust in de haven': De ontwikkeling van de arbeidsverhoudingen in de Rotterdamse haven 1945-1965*. Amsterdam: IISG, 1988.
Nijkamp, Bert. "Herinneringen aan het openbaar autobusvervoer in Zuid-Limburg: De jaren na de Tweede Wereldoorlog." *Het Land van Herle* 52 (2002): 1-13.
Nikolaeva, Anna, Peter Adey, Tim Cresswell, Jane Yeonjae Lee, Andre Nóvoa, and Cristina Temenos. "Commoning Mobility: Towards a New Politics of Mobility Transitions." *Transactions* 44, no. 2 (2019): 346-360.
Nite, Dhiraj. "Employee Benefits, Migration and Social Struggles: An Indian Coalfield, 1895-1970." *Labor History* 60, no. 4 (2019): 372-391.
Norton, Peter. *Fighting Traffic: The Dawn of the Motor Age in the American City*. Cambridge, MA: MIT Press, 2008.
———. "Four Paradigms: Traffic Safety in the Twentieth-Century United States." *Technology and Culture* 56, no. 2 (2015): 319-334.
———. "Jaywalking and the Invention of the Motor Age Street." *Technology and Culture* 48, no. 2 (2007): 331-359.
———. "Urban Transport and Mobility in *Technology and Culture*." *Technology and Culture* 61, no. 4 (2020): 1197-1211.

Oldenziel, Ruth, and Adri Albert de la Bruhèze. "Contested Spaces: Bicycle Lanes in Urban Europe, 1900-1995." *Transfers* 1, no. 2 (2011): 29-49.
Oldenziel, Ruth, Martin Emanuel, Adri Albert de la Bruhèze, and Frank Veraart. *Cycling Cities: The European Experience*. Eindhoven: Foundation for the History of Technology, 2016.

Oldenziel, Ruth, Mikael Hård. *Consumers, Users, Rebels: The People who Shaped Europe.* London: Palgrave, 2013.

Oldenziel, Ruth, and Milena Veenis. "Het huishouden tussen droom en werkelijkheid: oorlogseconomie in vredestijd, 1945-1963." In *Techniek in Nederland in de twintigste eeuw. Deel 4: Huishoudtechnologie, medische techniek*, edited by Johan Schot, Harry Lintsen, Arie Rip, and Adri Albert de la Bruhèze, 103-131. Eindhoven/Zutphen: Foundation for the History of Technology/Walburgpers, 2001.

Oldenziel, Ruth, and Heike Weber. "Introduction: Reconsidering Recycling." *Contemporary European History* 22, no. 3 (2013): 347-370.

Otten, Ad, and Elisabeth Klijn. *Philips' woningbouw 1900-1990: fundament van Woningstichting Hertog Hendrik van Lotharingen.* Zaltbommel: Europese Bibliotheek, 1991.

Ouweneel, Marieke. "Het belang van de Molukse wijk in Nederland." MSc Thesis, Utrecht University, 2011.

Overbosch, S. "Stopzetting arbeidersvervoer Enschede-Glanerbrug." *Op de Rails*, no. 4 (1951): 25-26.

Patton, Randall. "Textile Organizing in a Sunbelt South Community: Northwest Georgia's Carpet Industry in the Early 1960s." *Labor History* 39, no. 3 (1998): 291-309.

Peet, Jan. "Werken en leren 1948-1973." In *Vertrouwde patronen, nieuwe dromen: Nederland naar een modern industriële samenleving 1948-1973*, edited by Hélène Vossen, Marjan Schwegman and Peter Wester, 122-127. IJsselstein: VGN, 1992.

Peijs, R. "De landbouw in een stroomversnelling: Aspecten van de agrarische geschiedenis van Nederland 1945-1985." In *Vertrouwde patronen, nieuwe dromen: Nederland naar een modern industriële samenleving 1948-1973*, edited by Hélène Vossen, Marjan Schwegman and Peter Wester, 15-22. IJsselstein: VGN, 1992.

Pooley, Colin. "Mobility, Transport and Social Inclusion: Lessons from History." *Social Inclusion* 4, no. 3 (2016): 100-109.

———. "Balancing Social Justice and Environmental Justice: Mobility Inequalities in Britain Since Circa 1900." In *Transport Policy: Learning Lessons From History*, edited by Colin Divall, Julian Hine and Colin Pooley, 47-64. Farnham: Ashgate, 2016.

Pooley, Colin, and Jean Turnbull. "The Journey to Work: A Century of Change." *Area* 31, no. 3 (1999): 281-292.

———. "Modal Choice and Modal Change: The Journey to Work in Britain Since 1890." *Journal of Transport Geography* 8, no. 1 (2000): 11-24.

Post, Sjoerd. *Anak mas – Gouden kind; Molukkers in Nederland.* Groningen: Uitgeverij Noordboek, 2001.

Pot, Frank. *Zeggenschap over beloningssystemen 1850-1987.* Leiden: NIPG/TNO, 1988.

Ramirez, Bruno. "The Working-Class Struggle Against the Crisis: Self-Reduction of Prices in Italy." *Radical America* 10, no. 4 (1975): 143-150.

Rammler, Stephan. "The Wahlverwandtschaft of Modernity and Mobility." In *Tracing Mobilities: Towards a Cosmopolitan Perspective*, edited by Weert Canzler, Vincent Kaufmann and Sven Kesselring, 57-75. Hampshire/ Burlington: Ashgate, 2008.

Reijrink, Lonneke. *De betekenis van het begrip passende arbeid voor werklozen, kortdurende en langdurige arbeidsongeschikten.* MA Thesis, University of Tilburg, 2010.

Rietveld, Jos. "Fietsgebruik in Nederland: 1880-1996." De Oude Fiets. Accessed October 28, 2021. https://www.oudefiets.nl/2-algemeen/92-fietsgebruik-in-nederland-1880-1996.

———. "Fongers: Geschiedenis." Fongers. Accessed June 16, 2021. http://www.rijwiel.net/fongersn.htm.

Salon, Deborah, and Sumila Gulyani. "Mobility, Poverty, and Gender: Travel 'Choices' of Slum Residents in Nairobi, Kenya." *Transport Reviews* 30, no. 5 (2010): 641-657.
Schaechterle, Karl-Heinz. "Verkehrsuntersuchung N.V. Philips Eindhoven." Ulm: Technischen Hochschule Stuttgart/ Ingenieursbüro Professor Dr. Ing. N.E. Feuchtinger, 1961.
Schipper, Frank, Martin Emanuel, and Ruth Oldenziel. "Introduction: Historicizing Sustainable Urban Mobility." In *A U-Turn to the Future: Sustainable Urban Mobility since 1850*, edited by Martin Emanuel, Frank Schipper and Ruth Oldenziel, 1-26. New York: Berghahn Books, 2020.
Schivelbusch, Wolfgang. *The Railway Journey: The Industrialization of Time and Space in the Nineteenth Century.* Oakland, California: University of California Press, 2014 [1977].
Schot, Johan, Harry Lintsen, and Arie Rip. "Betwiste modernisering." In *Techniek in Nederland in de twintigste eeuw. Deel 1: Techniek in ontwikkeling, waterstaat, kantoor en informatietechnologie*, edited by Johan Schot, Harry Lintsen, Arie Rip, and Adri Albert de la Bruhèze, 19-36. Eindhoven/Zutphen: Foundation for the History of Technology/Walburgpers, 1998.
Schott, Dieter. "Suburbanizing the Masses for Profit or Welfare: Conflict and Cooperation Between Private and Municipal Interests in German Cities, 1890-1914." In *Suburbanizing the Masses: Public Transport and Urban Development in Historical Perspective*, edited by Colin Divall and Winstan Bond, 79-100. London/New York: Routledge, 2003.
Schrage, Peter, Erik Nijhof, and Piet Wielsma. "Inkomensontwikkeling van werkenden en werklozen in Nederland (1913-1939)." *Tijdschrift voor Sociale Geschiedenis* 14 (1989): 347-394.
Schumacher, Erik. *1942: Oorlog op alle fronten*. Series Leven in bezet Nederland. Edited by Erik Somers and René Kok. Houten/Amsterdam: Spectrum/NIOD, 2017.
Schuyt, Kees, and Ed Taverne. *1950: Welvaart in zwart-wit.* The Hague: Sdu Uitgevers, 2000.
Scott, Peter. "Women, Other 'Fresh Workers,' and the New Manufacturing Workforce of Interwar Britain." *The International Review of Social History* 45, no. 3 (2000): 449-474.
Sheller, Mimi. *Mobility Justice: The Politics of Movement in an Age of Extremes.* London/New York: Verso, 2018.
———. "Theorising Mobility Justice." *Tempo Social* 30, no. 2 (2018): 17-34.
Sheller, Mimi, and John Urry. "The New Mobilities Paradigm." *Environment and Planning A* 38, no. 2 (2006): 207-226.
Shove, Elizabeth. "The Shadowy Side of Innovation: Unmaking and Sustainability." *Technology Analysis & Strategic Management* 24, no. 4 (April 2012 2012): 363-375.
Spek, Jeroen van der. "Een eeuw lonen en prijzen (1870-1970)." *Sociale Maandstatistiek* 19 (1971): 418-426.
Staal, Peter-Eloy. *Automobilisme in Nederland: Een geschiedenis van gebruik, misbruik en nut.* Zutphen: Walburgpers, 2003.
Stadler, Alfred. "A Long Way to Walk: Bus Boycotts in Alexandra, 1940-1945." Paper presented at the *African Studies Seminar*, University of the Witwatersrand, Johannesburg, South Africa, 1979.
Stoffers, Manuel. "Fietsen in de Nederlandse bergen: Revisited." *SCHL Jaarboek* 61 (2016): 33-56.
Stoler, Ann Laura. *Along the Archival Grain: Epistemic Anxieties and Colonial Common Sense.* Princeton: Princeton University Press, 2009.
Stoop, Sjef. *De sociale fabriek: Sociale politiek bij Philips Eindhoven, Bayer Leverkusen en Hoogovens IJmuiden.* Utrecht: Stenfert Kroese, 1992.

Thompson, Edward Palmer. *The Making of the English Working Class.* London: Penguin, 2013 [1963].
———. "Time, Work-Discipline, and Industrial Capitalism." *Past & Present*, no. 38 (1967): 56-97.
Tjong Tjin Tai, Sue-Yen, Frank Veraart, and Mila Davids. "How the Netherlands became a bicycle nation: Users, firms and intermediaries, 1860–1940." *Business History* 57, no. 2 (2015): 257-289.
Tóth, Katalin. "Budapest: Reviving the Bicycle Lifestyle." In *Cycling Cities: The European Experience*, edited by Ruth Oldenziel, Martin Emanuel, Adri Albert de la Bruhèze, and Frank Veraart, 161-171. Eindhoven: Foundation for the History of Technology, 2016.

Urry, John. *Mobilities*. Cambridge/Malden: Polity, 2007.
———. "The 'System' of Automobility." *Theory, Culture & Society* 21, no. 4/5 (2004): 25-39.

Valk, Arnold van der. "Ruimtelijke ordening in Nederland 1948-1973." In *Vertrouwde patronen, nieuwe dromen: Nederland naar een modern industriële samenleving 1948-1973*, edited by Hélène Vossen, Marjan Schwegman, and Peter Wester, 128-133. IJsselstein: VGN, 1992.
Valk, Ineke van der. *Harde werkers: Migranten van het eerste uur langs Rijn en Lek, 1945-1985*. Zutphen: Walburgpers, 2009.
Vanoutrive, Thomas. "Commuting, Spatial Mismatch, and Transport Demand Management: The Case of Gateways." *Case Studies on Transport Policy* 7, no. 2 (2019): 489-496.
———. "Minder inkomen, dus minder mobiel, dus minder kansen." In *Armoede en Sociale Uitsluiting: Jaarboek 2018*, edited by Jill Coene, Peter Raeymaeckers, Bernard Hubeau, Tim Goedemé, Roy Remmen and An Van Haarlem, 277-290. Leuven/The Hague: Acco, 2018.
Veenman, Justus. *De sociale integratie van Molukkers*. Lelystad: Koninklijke Vermande, 1984.
Veraart, Frank. "Cycling in Numbers: 1920-2015." In *Cycling Cities: The European Experience*, edited by Ruth Oldenziel, Martin Emanuel, Adri Albert de la Bruhèze, and Frank Veraart, 200-229. Eindhoven: Foundation for the History of Technology, 2016.
Veraart, Frank, Martin Emanuel, and Ruth Oldenziel. "Eindhoven: Engineering a Path for Bikes?." In *Cycling Cities: The European Experience*, edited by Ruth Oldenziel, Martin Emanuel, Adri Albert de la Bruhèze, and Frank Veraart, 53-63. Eindhoven: Foundation for the History of Technology, 2016.
Veraart, Frank, Stijn Knuts, and Pascal Delheye. "Antwerp: Cycling Claims a Comeback." In *Cycling Cities: The European Experience*, edited by Ruth Oldenziel, Martin Emanuel, Adri Albert de la Bruhèze, and Frank Veraart, 89-99. Eindhoven: Foundation for the History of Technology, 2016.
Veraart, Frank, and Manuel Stoffers. "Southeast Limburg: Cycling Goes Downhill." In *Cycling Cities: The European Experience*, edited by Ruth Oldenziel, Martin Emanuel, Adri Albert de la Bruhèze, and Frank Veraart, 65-75. Eindhoven: Foundation for the History of Technology, 2016.
Verlaan, Tim. "Mobilization of the Masses: Dutch Planners, Local Politics, and the Threat of the Motor Age 1960-1980." *Journal of Urban History* 47, no. 1 (2019): 136-156.
Villeneuve, Dominic, and Vincent Kaufmann. "Exploring the Causes of Social Exclusion Related to Mobility for Non-Motorized Households." *Transportation Research Board* 2674, no. 8 (2020): 911-920.
Vleuten, Erik van der. "History and Technology in an Age of 'Grand Challenges': Raising Questions." *Technology and Culture* 61, no. 1 (2020): 260-271.
Vleuten, Erik van der, Ruth Oldenziel, and Mila Davids. *Engineering the Future, Understanding the Past: A Social History of Technology*. Amsterdam: Amsterdam University Press, 2017.
Vogel, Jaap. *Nabije vreemden: Een eeuw wonen en samenleven*. Cultuur en Migratie in Nederland. The Hague: SDU Uitgevers, 2005.
Vos, Nick. *De rauwe wet van vraag en aanbod: Arbeidsverhoudingen in de Twents-Gelderse textielindustrie, 1945 tot 1949*. PhD Dissertation, University of Amsterdam, 2011.

Waarden, Frans van. "Organisatie, arbeid en ondernemersbeleid in de twentse katoenindustrie." In *Ter Elfder Ure* 33, edited by Hugues Boekraad, Gabriël van den Brink, Henk Hoeks and Frits van Wel, 157-200. Nijmegen: SUN, 1983.
Walks, Alan. "Driving Cities: Automobility, Neoliberalism, and Urban Transformation." In *The Urban Political Economy and Ecology of Automobility: Driving Cities, Driving Inequality, Driving Politics*, edited by Alan Walks, 4-20. Studies in Urbanism and the City. London/New York: Routledge, 2015.

Weber, Alfred. *The Theory of Location of Industries.* Chicago: University of Chicago Press, 1929 [1909].

Weber, Max. *Economy and Society: An Outline of Interpretive Sociology.* Berkeley: University of California Press, 1978.

Westhoff, J.Th. *De directe mogelijkheden der werkverschaffing bij de werkloosheidsbestrijding.* Zwolle: J.J. Tijl, 1938.

Willmott, Peter. "Car Ownership in the London Metropolitan Region." *Greater London Council Intelligence Unit Quarterly Bulletin* no. 23 (1973): 5-19.

Wolters, Luc. *Limburgse tijdreis met bus en tram: 90 jaar gemeentelijke activiteit in vervoermobiliteit.* Maastricht/Sittard: Gebaltram, 2010.

Wouden, Ries van der. "Succes of falen? Een halve eeuw verstedelijkingsbeleid in Nederland." *Ruimte & Maatschappij* 8, no. 1 (September 2016): 6-26.

Zwemer, Jan. "De pendelarbeiders van Tholen en Sint-Philipsland." In *Zeeland 1950-1965*, edited by Jan Zwemer, 401-428. Vlissingen: Uitgever Den Boer/De Ruiter, 2005.

Index

For Product Safety Concerns and Information please contact our EU representative GPSR@taylorandfrancis.com Taylor & Francis Verlag GmbH, Kaufingerstraße 24, 80331 München, Germany

Batch number: 10722100

Printed by Printforce, the Netherlands